**World Steel
in the 1980s:
A Case of Survival**

World Steel in the 1980s

A Case of Survival

William T. Hogan, S.J.
Fordham University

LexingtonBooks
D.C. Heath and Company
Lexington, Massachusetts
Toronto

Library of Congress Cataloging in Publication Data

Hogan, William Thomas, 1919–
  World steel in the 1980s.

  Includes bibliographical references and index.
  1. Steel industry and trade. I. Title.
HD9510.5.H57   338.4'7669142   75–41587
ISBN 0–669–00465–0   AACR2

Copyright © 1983 by D.C. Heath and Company

Third printing, September 1983.

Published simultaneously in Canada

Printed in the United States of America

International Standard Book Number: 0–669–00465–0

Library of Congress Catalog Card Number: 75–41587

*To Joseph Gerney,
industrialist and friend,
who contributed much to the development
of the world steel industry.*

Contents

Tables   xi

Preface   xv

Chapter 1   **The Current Steel-Industry Problem**   1

Growth   2
Ownership   4
International Trade   6
Raw Materials   7
Technology   10

Chapter 2   **The Steel Boom Collapses**   15

The Boom-Years Euphoria   15
The Collapse of the Boom   16

Chapter 3   **Western Europe Faces the Crisis**   19

EEC   19
Price Stability   23
Long-Range Rationalization   23
Raw Materials in the EEC   32
Technology   41
Ownership   45
EEC Trade   47
The Basic EEC Problem and Future Outlook   49
Other Western European Countries   51

Chapter 4   **Japan Seeks New Markets**   63

Growth   64
Raw Materials   67
Technology   76
Trade   81
Ownership   87
Future Developments   87

Chapter 5   The United States Faces Critical Decisions   91

Growth   92
Profits   96
Raw Materials   97
Technology   105
Ownership   113
Trade   113
The Future of the U.S. Steel Industry   119

Chapter 6   Prospects for Canada, Australia, South Africa,
and New Zealand   137

Canada   137
Australia   144
South Africa   148
New Zealand   150

Chapter 7   The Third World Moves toward Self-Sufficiency   153

China   154
China and Japan   159
Brazil   161
India   163
South Korea   165
Mexico   166
Taiwan   168
Argentina   170
Venezuela   172
Other Third World Countries   174
Future Growth in the Third World   175

Chapter 8   The Soviet Bloc Faces a Declining Growth Rate   183

Future Prospects   187

Chapter 9   International Trade   191
Japanese Export Activity   200
The 1977 Crisis   201
Future of Steel Trade   206
Devices Restricting Trade   206
Japan Must Export   208
Western European Exports   211
The United States as an Import Market   211
Other Third World Export Activity   213
International Steel Committee   213

Chapter 10   **A Case of Survival**   219

Future Growth Limited to the Third World   219
Raw Materials Remain Abundant   223
New Technology Essential but Costly   224
Shift to Public Ownership Continues   227
International Trade Problems to Recur   230
Access to the U.S. Market   236

Bibliography   241

Index   250

About the Author   272

Tables

2-1   Total New Steelmaking Capacity Planned for the
Non-Communist World: 1974   17

3-1   Production of Ore in the EEC for Selected Years:
1950–1979   33

3-2   EEC Imports of Iron Ore, by Country of Origin: 1979   36

3-3   EEC Crude-Steel Production, by Process: 1980   42

3-4   Growth of Oxygen Steelmaking in the EEC for
Selected Years: 1968–1980   42

3-5   EEC Continuously Cast Steel Output: 1969, 1974,
1975, 1979, and 1980   44

3-6   Steel Capacities Owned or Controlled by Government
in the EEC: 1965 and 1981   47

3-7   Steel Imports and Exports for EEC Countries:
1966–1974   48

3-8   Crude-Steel Production of Non-EEC Western
European Countries: 1965 and 1980   52

3-9   Crude-Steel Production, by Process, for Non-EEC
Western European Countries: 1980   53

3-10   Steel Continuously Cast, by Non-EEC Western
European Countries: 1980   53

3-11   Spanish Raw-Steel Production for Selected Years:
1946–1980   55

4-1   Major Sources of Japanese Ore Imports for Selected
Years: 1962–1979   70

4-2   Imports and Domestic Production of Japanese
Coking Coal: 1960–1980   71

4–3   Japanese Imports of Coking Coal for Selected Years: 1962–1980   72

4–4   Japanese Imports of Scrap for Selected Years: 1960–1980   74

4–5   Continuously Cast Steel Output in Japan: 1969–1980   78

4–6   Japanese Exports of Steel Products: 1960–1980   82

4–7   Japanese Raw-Steel Production and Exports: 1960–1980   82

4–8   Steel Exports of Newly Industrialized Countries: 1973–1979   86

5–1   North American Pelletizing Capacity   99

5–2   U.S. Iron-Ore Consumption and Imports, by Principal Countries of Origin: 1970–1980   100

5–3   Tonnage of Continuously Cast Steel in Selected Countries: 1972–1980   109

5–4   Imports and Exports of Steel Products for the United States: 1960–1980   115

5–5   U.S. Automobile Production, Total Steel Shipments, and Sheet and Strip Shipments to the Automotive Industry: 1955–1980   125

6–1   Steel Producers in Canada   139

6–2   Canadian Imports and Exports of Steel Products: 1971–1980   142

6–3   Exports and Imports of Australian Steel Products: 1970–1979   145

6–4   Exports of Australian Iron Ore: 1967–1980   147

7–1   Raw-Steel Output for the Leading Third World Producers: 1970, 1974, and 1980   155

7-2   Japanese Steel Exports to Principal Consumers:
        1970–1980   160

7-3   Brazilian Imports and Exports of Steel and their U.S.
        Dollar Value: 1974–1980   163

8-1   Growth of Soviet Bloc Steel Production: 1966–1980   184

8-2   Crude-Steel Production, by Process, for Eastern
        Bloc Countries: 1974 and 1980   185

9-1   Exports for the European Common Market
        Countries, Japan, and the United States: 1967–1974   193

9-2   Imports of Steel for Selected Third World Countries:
        1967–1974   195

9-3   Indexes for Steel Production for the United States,
        Japan, and the EEC Countries: 1964–1974   196

Preface

Although this book considers the major influences likely to shape the world steel industry over the balance of this decade, its principal focus is on the years through 1985. The basic reason for this is that the uncertain conditions that exist in the world steel industry today have made it virtually impossible for steel companies in the industrialized world, as well as those in the Third World, to formulate anything but tentative plans beyond 1985. This is particularly true of Western Europe and the United States where the steel industry's capacity is shrinking noticeably. The Japanese industry, which is geared to the world market, also has problems because of the uncertain nature of world steel demand. Japan is facing increasing competition from the facilities that have been and are being installed in the Third World, so any expansion of the Japanese steel industry for the remainder of the decade is extremely doubtful.

Unlike the industrialized countries, the Third World in the late 1970s and early 1980s embarked on a program of expansion. However, with the exception of one or two countries, most notably South Korea, these plans have not materialized according to the timetables that were set. A number of Third World countries have had to reassess their respective situations and retrench.

This book's subtitle, "A Case of Survival," directs attention to the question of whether or not the steel industry will survive as it is presently organized. There is no question that the steel industry on a global basis will survive and continue to meet most of the demands made upon it. However, the problem is whether or not the existing companies will survive as they are today in terms of size, products produced, and type of ownership, that is, public or private. Severe competition, not only within the steel industry itself but also from other materials, has rendered many of the companies profitless; in fact, in the past six years, substantial losses have been sustained. If this condition is not reversed within the next two to three years, a number of these companies will either disappear or be completely reorganized, while a number of privately owned companies could be taken over by their governments. It is in this sense that the term *survival* is used in the book's title.

In treating the past, present, and future developments on a country-by-country basis, it was not the author's intention to present a full development of each country's steel industry, but rather to indicate what has happened and what could happen in terms of growth, ownership, the use of technology, the availability of raw materials, and the country's position in international steel trade. For the sake of continuity in individual chapters, certain subjects are discussed in more than one location throughout the text.

Much of the information included in this book concerning the future plans of steel producers was obtained from personal contact with executives in virtually all of the steel companies in the non-Communist world during the past two years. The treatment of the USSR is more limited than the rest of the world for two reasons: (1) the USSR is not a factor in steel exports outside of its own orbit, and (2) comprehensive information concerning future plans is not readily available.

In the discussions on international trade in steel, the United States has received more attention than the other steel-producing countries. This seemed to be justified since the United States is not only the largest net importer of steel by far, but also, almost every steel-producing country, with the exception of the USSR and some of its Eastern European satellites, has directed large tonnages of steel to the U.S. market.

Unless otherwise noted, the measurement used throughout this book is a metric ton.

1

The Current Steel-Industry Problem

The world steel industry of the 1980s faces severe problems that must be solved if it is to survive as a productive and profitable component of the world's industrial economy. On a global basis, the industry must contend with uncertainty over future market growth, difficulties with international trade, the need to open new sources of raw materials, as well as the necessity to develop and apply new technology. Likewise, the ownership of the industry, whether public or private, could change significantly in the years ahead.

The problems confronting the industry vary from country to country, both in kind and intensity. However, the global nature of the industry is such that difficulties in one region have a definite impact on steel activity in other areas. Developments during the years immediately ahead will determine whether world steel can survive as a source of real economic progress and prosperity or whether it will decline as an ailing, subsidy-dependent industry.

Since 1974, the height of the steel boom, the large, industrialized countries including Japan, the United States, and those in Western Europe have suffered sharp cuts in steel production and reductions in earnings that, in some cases, have become severe losses. These financial reverses, in a number of instances, have led to plant closures and bankruptcies. Further, decreased demand and low rates of operation have resulted in a series of decisions, curtailing expansion and placing emphasis on the modernization of existing facilities, the elimination of those that are obsolete, and the conservation of energy.

Without question, the most critical region in the steel industry during the second half of the 1970s was and still is Western Europe. Sharp drops in production, coupled with a crumbling price structure, have resulted in losses of hundreds of millions of dollars for most of the integrated companies, forcing a number of them to look for government help to avoid bankruptcy. The crisis, which began in 1975, not only has persisted to the present day but has definitely worsened. The steel industry in Western Europe is fighting for survival.

In the United States, some 12.5 million tons of integrated steel capacity have been taken out of operation since 1977, as the industry trims down to its efficient capability.[1] Expansion plans have been virtually eliminated, and capital funds are being used for modernization, diversification, and pollution containment.

The Japanese industry has suffered a considerable drop in production and shipments, as well as in export tonnage to the rest of the world. Because of operating rates between 70 and 75 percent of capacity, expansion has been halted and investments have been concentrated on modernization of facilities, cost reduction, and energy conservation.

An exception to the downward trend in the industrialized world is Canada, where steel production and earnings have surpassed the 1974 records by significant amounts.

The Third World, in contrast to most of the industrialized world, has expanded its production by 67 percent between 1974 and 1980, as output rose from 60.3 million to 100.8 million tons during that period. Nine countries in the developing world are capable of producing 2 million or more tons of steel, and most of these have ambitious plans for expansion.

The current crisis in steel, which has persisted for several years, was brought on by developments in previous decades, particularly the 1960s and 1970s. During those years, a number of social and political influences led to structural changes affecting the growth and ownership of the world steel industry. Further, a rapid expansion in international trade had a significant effect on steel prices, profits, and employment. There were also widespread applications of new technology and shifts in the sources of raw materials. These changes profoundly influenced the industry and will continue to have a decided impact on it during the 1980s.

Some comments on the changes giving rise to current problems in a world context follow, and in the remaining chapters, they are more fully analyzed on a regional basis.

Growth

During the 1960s and early 1970s, in most parts of the world, steel-industry output, triggered by a dramatic increase in the markets for steel, grew steadily. Global production in 1960 was 340 million tons, rising to 595 million in 1970, and culminating in the record output up to that time of 709 million tons in 1974.[2]

This huge increase in production over so short a period was in response to the growing demand for products made of steel, particularly after the mid-1950s, when the world began to emerge from a war that had inflicted such widespread destruction that entire nations had to be rebuilt. With the completion of the fundamental repairs to various war-torn economies, the desire for an improved life-style among the Western Europeans and Japanese resulted in an ever-increasing demand for consumer durable goods. One of the leading influences in the steel industry's expansion was the tremendous increase throughout the 1960s and 1970s in the demand for automobiles. For the first time, a large segment of the population in Western Europe and Japan bought automobiles. When compared with the number of motor vehicles in those areas before World War II, and immediately thereafter, this increase in the postwar period was truly phenomenal. In 1938, for example, automobile production in Japan was a mere 24,000 units, and only 128,000 units were registered, or one for 623 persons. In Germany, production was 352,000 units, with 950,000 registered, or one vehicle for every 137 individuals. In the postwar period these numbers changed radically as automobile production in Japan and Western Europe became a major factor in the economy.

In 1950, world production of motor vehicles, including passenger cars and trucks, was 10.6 million units, with some 70.4 million registered and in operation. By 1960, production had increased to 16.4 million units; in 1970, it was 29.7 million units; and in the next three years, an abrupt rise in production brought vehicle output to 39.2 million units in 1973. In the following years, there was a significant drop to 33.3 million in 1975, followed by a sharp recovery that raised production to 42.5 million units in 1978. In Western Europe, the number of people per vehicle had fallen to an average of three in 1978, while in Japan it was reduced to 3.4.[3]

These developments in automobile production on a worldwide basis spurred an increase in the demand for steel. The motor-car industry was the principal market for steel in the United States, while in Japan and Western Europe it ranked either second or third depending on the year.

Worldwide automobile registrations, which indicate the number of vehicles in operation, rose from 70.4 million in 1950 to 127.0 million in 1960 to 206.4 million in 1970 to 396.0 million in 1979. The operation of this vastly increased number of vehicles brought a correspondingly rapid increase in the demand for petroleum products. World crude-oil production in 1950 was 3.8 billion barrels, and by 1978, it had risen to 22.2 billion barrels.[4] Further, the location of most of the known oil reserves shifted to the Middle East and to other areas, many of which were quite distant from the point of use, requiring extensive large-tanker and pipeline construction to carry the oil over long distances. The size of the oil tanker grew from 50,000 deadweight tons in 1950 to 250,000 deadweight tons in the next fifteen years, while a few vessels were in the 400,000–500,000-deadweight-ton range.

Shipbuilding was and still is one of the leading markets for Japanese steel. Until recently, shipbuilding took more steel tonnage than the Japanese automobile industry. Pipelines, hundreds of miles in length, also consumed large quantities of steel.

The impact of automobile growth was felt in the oil industry that, in turn, called for more steel in drilling, refining, and transportation. The more than fivefold increase in automobile registrations over a twenty-eight-year period stimulated an almost sixfold increase in petroleum production, a very large increase in pipeline and ship construction and, without question, was a major influence on the fourfold increase in steel production during the same period. An even greater influence, but more difficult to pinpoint, was the investments made in the capital-goods sectors of the Western European, Japanese, and North American economies.

The growth in steel demand in the 1960s was unusual insofar as it was characterized by a steady increase in output during the decade. On a worldwide basis, no downward cyclical changes occurred in steel production; in fact, in the entire post-World War II period, up to and including 1974, in only three years did production decline below the previous year, and in each of these instances, on a worldwide basis, the decline was minimal.

In 1954, world production dropped to 224 million tons from 235 million; in 1958, it fell to 274 million from 292 million tons. The last decrease in 1971 was minimal, as output fell to 582 million tons from 595 million of the previous year, an extremely small decline.

The United States experienced cyclical fluctuations during this period, but the rest of the world did not. In 1950, U.S. production stood at 88 million tons increasing to 102 million in 1953, dropping back to 80 million in 1954, rising to 106 million in 1955, and dropping to 77 million in 1958. It rose to 128 million in 1969, fell to 109 million in 1971, and reached its all-time peak of 137 million in 1973. By contrast, in the entire period between 1950 and 1973, the Japanese witnessed only one year with a significant decline in production, that was 1971 when output fell by almost 5 million tons to 88.6 million from 93.3 million for the previous year. The other year in which Japan had a drop in production, an almost insignificant drop, was 1962, when output fell by 800,000 tons below the previous year. This record of continued growth inspired the steel industries of the world, outside of the United States, to install more and more capacity.

In the early 1970s, growth projections were highly optimistic; however, as 1975 brought an end to the boom and produced a recession that has lasted for several years, many steelmakers changed their minds about the industry's growth potential. Views continue to differ in various parts of the world. The industrialized countries with their large capacities tend to be pessimistic, while the Third World, which is anxious to grow, is much more optimistic about the future.

Ownership

A significant change took place in the ownership of the steel industry during the postwar period. In 1950, 77 percent of the world's steel output was in

private hands, with the balance of 23 percent under government ownership.[5] The growth of the industry in the USSR and Eastern Europe, as well
as the complete or partial nationalization of steel companies in other countries, changed the complexion of the industry's ownership during the 1960s
and 1970s. By 1980, approximately 60 percent of the world's steel production came from government-owned or heavily subsidized mills, with 40
percent from privately owned facilities. The increase in the USSR's production was one of the most significant factors in this change. In 1950, raw-
steel output for the USSR was 27 million tons; by 1980, it was 147.9 million
tons, having peaked in 1978 at 151.4 million tons. In Eastern Europe, where
the industry is completely state owned, production in 1950 was slightly less
than 8 million tons; in 1980, it was approximately 61.2 million tons.

Much of the Third World capacity installed during the postwar period
is government owned. This applies to the entire 37 million tons in China, the
5 million to 6 million tons in North Korea, and major portions of the steel
industries of Brazil, South Korea, Taiwan, Mexico, Venezuela, and Argentina.

In the industrialized world, nationalization moved large companies
from the private to the public sector. The largest tonnage to be nationalized
was that of British Steel Corporation (BSC), where some 28 million tons
passed to government ownership as a result of the 1967 Nationalization
Act.[6] In France and Belgium, as well as Italy, major portions of the industry
are either nationalized or heavily subsidized by the government and, thus,
subject to government control. In the industrialized world, the principal
areas of private ownership are the United States, Japan, Canada, West Germany, Luxembourg, Holland, and Australia.

The shift in ownership has brought with it some thorny problems, particularly when the worldwide steel industry is in a depressed condition. The
steel industry in most countries is a large employer, and the principal objective of a government-owned company is to maintain employment. In order
to do this, operations must be sustained at a high level, and this, in a
depressed market, almost always necessitates a reduction in the price of
steel, particularly steel that is exported and that must compete in other
markets. Thus, these companies have incurred severe losses that have been
absorbed in one way or another by their governments.

Privately owned companies must depend on their own resources and
cannot operate for any extended period of time with large deficits. Their
price structure, generally speaking, in the long run must cover their total
costs of production, and in this respect, they have been placed in an unfavorable competitive position in the past few years. Solutions to the problem, wherever private and public companies come into conflict because of
price competition, have involved government intervention in one form or
another to restrict tonnage output and stabilize the price structure.

The conflict between privately and publicly owned companies has been plaguing Western Europe for most of the time since 1975. There has been a controversy in the Common Market between privately owned and publicly owned companies concerning subsidies that place the privately owned companies at a competitive disadvantage in pricing their products.

This is one of the most serious problems facing the steel industry in the 1980s, and hopefully, a solution can be found.

International Trade

International trade in steel has grown dramatically in the post–World War II period as the industries in Western Europe and Japan expanded far beyond their prewar levels. With added capacities, these countries not only were able to supply their own needs but also had surplus steel to trade. In 1950, 11 percent of world-steel production found its way into international trade, as a total of 20.5 million tons of raw-steel equivalent was moved across international borders. By 1979, this increased to 25 percent of the world's steel production, or the raw-steel equivalent of 181.2 million tons, a ninefold rise.[7]

The United States, which was the dominant steel producer in the world during most of the postwar period, became the leading importer of steel in the 1960s and 1970s as other countries sought a share of the large, lucrative U.S. market.

The Japanese, newcomers to large-scale steel output, designated much of their increased steel capacity for international trade, with the United States becoming their principal outlet. During the last two decades, Japan has become the world's leading exporter of steel.

The growth of the steel industry in Western Europe, which resulted in a fivefold increase in output between 1947 and 1974, provided considerable tonnage for export from countries such as West Germany, France, Belgium, Luxembourg, and Italy.

Unlike a number of other commodities that are traded between the have and have-not nations, the major portion of international trade in steel takes place between steel-producing countries. In order to penetrate markets in a country that has considerable steel production, outside producers have found it necessary to offer price concessions to potential customers. During a period of normal or less than normal demand, import prices are usually below those charged by domestic producers. However, when demand increases, as it did in 1973 and 1974, the import prices increase sharply to take advantage of the market situation. Unfortunately, for the past seven years, world demand has been considerably below potential supply. Thus, exporters have had to offer lower prices to maintain their foreign markets.

The situation deteriorated to such an extent in 1977 that exporters to the United States were offering steel at prices $100 per ton below the U.S.

list price. Further, import volume was so great, it shattered the price structure in the United States and was in part responsible for sharp decreases in profits and some bankruptcies.

In Western Europe, virtually the same situation developed as steel producers in the Common Market offered substantial price concessions to customers throughout the European Economic Community (EEC) in order to obtain business. These concessions were almost always met by competing steel companies, with the result that the price structure was destroyed in 1977, and although a number of efforts have been made to restore it, until mid-1981 they had not been successful.

The desire of a number of major steel companies throughout the world to maintain employment in depressed periods led to lower prices for exported steel. This policy brought restrictions against imports to protect domestic steel companies and steel-industry employment. The restrictions began in 1968 when imports to the United States rose to a record 16.3 million tons, inspiring management and labor to appeal to Congress for temporary tariffs or quotas on steel imports. Wishing to avoid these restraints, the Western Europeans and Japanese offered voluntarily to limit their exports to the United States.

In 1972, the EEC placed quotas of 1.2 million tons per annum[8] on imports from the six Japanese integrated companies. In 1977, the United States instituted the trigger-price mechanism to increase the price of imports and, hopefully, to limit the tonnage imported.

The attempt by steel companies and governments to maintain employment in depressed periods, which has led to restrictions on steel trade, is intimately tied to the past growth of the steel industry, as well as its present ownership. The restrictions on international trade in steel, as indicated, have taken a number of forms and will almost certainly continue to be imposed for some time to come.

Raw Materials

The raw materials for steelmaking consist essentially of iron ore and coking coal, as well as some fluxing agent, usually limestone, which are fed into the blast furnace to make pig iron. Scrap, although not a raw material in the sense of iron ore and coal, is nevertheless the basic feed material for the electric furnace and, as such, might be considered a raw material.

Iron Ore

In the immediate postwar period, great concern existed among the steelmakers of the industrialized world about the future supply of iron ore. Deposits in the steel-producing countries were fast becoming depleted, and there was a considerable sense of urgency to replace them. Fortunately, huge rich reserves were discovered from the late 1940s through the 1960s in Venezuela, Brazil, Australia, Canada, and Africa so that iron ore is now in plentiful supply. Unfortunately, in many instances, the supply is far removed from the steel-producing centers and must be shipped over long distances. This is particularly true for Japan and Western Europe, while the United States is fortunate since most of its imports come from nearby Canada.

The development of the steel industry in Japan, where virtually no iron ore exists, and the need of other countries to import ore have greatly increased trade in this basic raw material. In 1950, total production in the world was 246 million tons, of which 40.5 million, or 16.5 percent, were shipped in international trade.[9] In 1974, a year of record steel production up to that time, 883.0 million tons of ore were mined, of which 402.0 million tons, or over 45 percent, were shipped in international trade. Thus, we find that, in a matter of twenty-five years, ore shipments in international trade far exceeded the total production at the beginning of the period. In 1979, 394.0 million tons out of a total production of 885.0 million tons of ore, or approximately 45 percent, were shipped in international trade.

The Japanese are principally responsible for the huge tonnages, importing as they do virtually 100 percent of their ore, that amounted to 133.7 million tons in 1980. This was a tenfold increase over 1960, when steel production of 22.0 million tons required ore imports of 13.4 million tons. The United States and Western Europe have also become large importers of ore since the end of World War II.

Iron ore is abundant throughout the world because, in addition to the new discoveries, which in some cases are in remote locations, large-scale programs have been undertaken to beneficiate low-grade-ore deposits in order to make them acceptable as a blast-furnace feed. This material, usually in the form of pellets or sinter, is now available in large tonnages.

One fact stands out in the matter of iron-ore supply: The discovery of new deposits in remote locations, as well as the developments in the beneficiation of low-grade ores that require large capital investments, have raised the costs of obtaining iron ore for the steel industry's blast furnaces. The price of ore will increase in the years ahead to cover these costs.

Coal

Metallurgical coal for the production of blast-furnace coke is by no means as abundant as iron ore. A number of countries such as Japan, Italy, France, Spain, South Korea, Brazil, and Argentina have very little of this resource and must import it. These areas are dependent for metallurgical coal principally on the United States, Canada, Australia, and Poland. At the height of the steel boom in 1973–1974, metallurgical coal was in tight supply, and spot purchases commanded prices of $100 per ton.

The supply of this material has not increased much through new discoveries; however, the technology of coke production has improved to a point at which inferior coals can be used. These developments include briquetting part of the charge, as well as the use of about 15 percent of what was formerly considered noncoking coal in the total blend of coals for the coke-oven charge. Such devices have by no means solved this problem but, to some extent, have alleviated it. For the foreseeable future, metallurgical coal will be in demand for the production of pig iron, which is essential in the basic-oxygen steelmaking process.

Scrap

Scrap will be in greater demand as electric-furnace capacity and output continue to expand. The electric-furnace process has grown rapidly in the last fifteen years to a point at which it now provides more than 160.0 million tons of the world's production, as opposed to less than 20.0 million tons in 1950. In most instances, electric furnaces use a 100 percent scrap charge, and the sheer increase in the capacity of these furnaces has placed a tremendous demand on the available supply of scrap.

The United States is the chief source of supply to the world, since it generates large quantities of obsolete metal. Upwards of two billion tons of iron and steel have gone into the economic structure of the country during the past century, and a certain percentage of this becomes available for processing into scrap each year. In Western Europe, West Germany, the United Kingdom, and France export significant quantities of scrap to their neighboring countries.

The future supply of scrap is somewhat clouded, first, by the projected increase in electric-furnace production; second, by the deterioration of scrap quality; and third, by the fact that continuous casting will reduce the amount of mill-revert scrap available.

A significant portion of the additional electric-furnace capacity will be devoted to making the higher grades of steel and, thus, operators will not be able to tolerate the inclusion of high percentages of residual elements in the scrap. It is hoped that the problem can be solved in the long run by careful selection of scrap or by adding direct-reduced iron (DRI) to the furnace charge. This material is not widely available and it will be some time before large quantities are produced.

Scrap will continue to be the principal feed material for electric furnaces during the 1980s, although with the growth of the electric furnace, the availability of high-quality scrap could be limited.

Technology

During the postwar period, a great number of technological developments have been made. Some of these can be considered basic breakthroughs in the steelmaking process, while others, although significant, are minor or represent the application to steel of technology developed in other areas—computers, for a prime example.

Among the basic breakthroughs, three stand out: (1) the oxygen converter, which has revolutionized the steelmaking process; (2) continuous casting, which has eliminated four steps from the conventional steelmaking process and has conserved substantial quantities of energy; and (3) the direct reduction of iron ore outside of the blast furnace, which makes it possible to produce iron without coke made from metallurgical coal.

The first process, oxygen steelmaking, has replaced the open hearth as the principal steelmaking facility and, by 1990, will be producing at least two-thirds of the world's total steel output. It is a pneumatic steelmaking technique in which commercially pure oxygen refines a charge of molten iron and recycled scrap held in a pear-shaped vessel or converter. The primary advantage of the oxygen process is its speed of operation that substantially shortens the required production time. Although a number of variations in oxygen steelmaking are currently in commercial use, the most common version is termed the *basic-oxygen* process; outside of the United States it is called the LD process. It introduces a top-mounted lance through the mouth of the vessel to inject oxygen downward onto the charge and routinely produces heats as large as 300 tons in forty-five-minute cycles, compared to more than six hours in the open hearth. The steel obtained is at least equivalent in every respect.

The second advance, continuous casting, unlike the oxygen converter, has witnessed a relatively slow acceptance. However, within the past three to four years, the rate of installation of these units has increased rapidly and will continue to do so for the rest of the decade. In 1980, the steel industry cast approximately 30 percent of world production, up from 14 percent in 1975. In terms of tonnage, the output of continuously cast steel rose from 85.0 million to 195.0 million tons over the five-year period.[10]

In continuous casting, molten steel tapped from the steelmaking furnace into a bottom-pour ladle is introduced into a tundish, or reservoir, and passed into the water-cooled mold of the casting machine, where it solidifies and emerges continuously in semifinished form. This eliminates the need to pour the steel into ingot molds, strip the molds from the ingots, place the ingots into soaking pits to develop an even temperature, and finally, roll the reheated ingots on a primary rolling mill to obtain the desired semifinished product. Without the many croppings involved in rolling ingots, continuous casting results in significant yield improvements. By eliminating ingot cooling and soaking-pit reheating, significant energy savings are also realized.

The third breakthrough, direct reduction, in terms of its output is still in its infancy. It replaces the traditional blast-furnace method for removing oxygen from iron ore to produce steelmaking iron. Direct reduction refers to any process using gas or solid fuels to remove the oxygen from the ore, bypassing the blast furnace. The process, which takes place in a shaft furnace or kiln, results in a metallurgic substance that usually has more than 90 percent iron content. The end product is referred to as direct-reduced iron (DRI). Compared to blast-furnace iron, DRI usually has a much lower carbon content (2 percent maximum), that is ideally suited to maintaining the melting action in the electric steelmaking furnace. It also contains 6 percent or more of unreduced or partially reduced iron oxides and noniron-bearing substances, normally referred to as gangue.

At the end of 1980, the world's rated direct-reduction capacity reached 15.5 million annual tons, up sharply from 11.3 million tons at year's end 1979. World DRI output in 1980 totaled a record 7.2 million tons, with the natural-gas-based Midrex and HYL processes accounting for 95 percent of the total. The leading producer nations were Mexico and Venezuela, with a combined output share of 42 percent, followed by Canada, Argentina, and the United States.[11] These five nations accounted for nearly three-fourths of 1980's total world production.

The most successful of the direct-reduction processes are based on natural gas, which is economically available in relatively few countries throughout the world. Efforts are under way to develop processes, using coal as a reducing agent in place of gas, that are economically and technologically sound. One such process is in operation in several locations around the world; however, it has had its problems. If success is achieved in using coal, the resultant process will have substantial advantages insofar as coal is much more abundantly available throughout the world than natural gas.

In those areas where gas is available, a number of direct-reduction units are currently either being installed or are in the planning stage. By the mid-1980s, production of direct-reduced material using gas as a reducing agent will probably be in excess of 20.0 million tons.

A number of other technological developments are in the experimental or pilot-plant stage. One of these is formed coke. This process permits the use of nonmetallurgical coal and also reduces air pollution. As yet, no large plants have been installed anywhere in the world. The largest producing unit is in Wyoming in the United States where 225 tons per day can be turned out. Many believe this process has great potential; however, it will take considerable capital investment and time before it becomes a large-scale method of producing coke.

Other highly significant technological developments include:

The beneficiation of low-grade ores and their transformation into pellets;

Desulfurization of iron in the ladle before use in the steelmaking furnaces;

Ultrahigh power (UHP) in the electric-furnace operation;

Argon oxygen decarburization (AOD) in the production of specialty steels;

Continuous annealing for sheet products;

Computer application to a number of steps in the steelmaking and finishing processes.

The basic problem with the widespread use of new technology is the capital cost of installing the facilities. For example, large-scale, continuous-casting units, with a capacity for 1.0–1.5 million tons annually, require capital investments in the range of $100 million. A new hot-strip mill involves an investment of $350–400 million.

The large amounts of capital needed to install equipment and the capital charges involved in operating it, including interest, amortization, and depreciation, very often neutralize some of the economic gains that the innovations provide. Consequently, decisions will have to be made in the next decade as to whether the new technology should be installed or existing facilities and technology upgraded wherever possible.

The steel industry is facing competition not only from within its own ranks but also from other materials such as aluminum, plastics, and cement. It should, whenever possible, employ the latest technology. Further, there is a need for more technological breakthroughs that will reduce capital costs and permit steel to contain its operating costs.

Notes

1. This figure is compiled from the announcements of the various companies that have shut down steelmaking capacity.

2. Figures on steel production for individual countries and the world are taken from International Iron and Steel Institute, *Statistical Reports* (Brussels), which are issued monthly and annually.

3. Statistics on automobile production and registration are taken from Automobile Manufacturers' Association, *Facts and Figures* (Detroit, selected years). The name of this association was changed in 1972 to Motor Vehicle Manufacturers Association of the United States.

4. *Basic Petroleum Data Book* (Washington, D.C.: American Petroleum Institute, 1979).

5. Percentages are derived from the production statistics of the companies, both privately and publicly held.

6. Figures are compiled from the individual capacities of the fourteen companies that made up BSC.

7. Statistics on international trade are taken from International Iron and Steel Institute, *World Steel Figures* (Brussels, 1981), p. 14.

8. This agreement was entered into in 1972 between the Japanese Ministry of International Trade and Industry and the European Coal and Steel Community (ECSC).

9. U.S. Department of the Interior, Bureau of Mines, *Minerals Yearbook,* vol. 3 (Washington, D.C.: various years).

10. International Iron and Steel Institute, *World Steel Figures,* p. 8.

11. William T. Hogan and Frank T. Koelble, *Direct Reduction as an Ironmaking Alternative in the United States* (Washington, D.C.: U.S. Department of Commerce, 1981), p. 14.

2 The Steel Boom Collapses

The problems facing the worldwide steel industry today have intensified during the past six years, although their roots go back much further. They were not foreseen as recently as the early 1970s, when a worldwide boom, resulting from a record aggregate demand from consuming industries, produced thoughts of continued growth and expansion.

In 1973, world raw-steel output reached a record 698.0 million tons, as the result of an unprecedented 11 percent gain (68.0 million tons) over the previous year. In 1974, a further increase in production established another record of 709.0 million tons. This gain was relatively modest because of the limitations on available production capacity and essential raw materials. Profits were high in Western Europe and Japan, while in the United States they reached a 2.5-billion dollar record.

The Boom-Years Euphoria

During the boom years, optimism among steelmakers was unbounded. This was evident at the Johannesburg Meeting of the International Iron and Steel Institute (IISI) in October 1973, when steelmen from thirty-eight countries talked about prosperity and expansion. The mood of the meeting can be summed up in the following comments of one of the speakers:

> The past 12 months have been eventful ones for the steel industry. Caught, as we are now, in the strong worldwide upsurge of demand for all kinds of steel products, the primary concern of all steel producers is to maximize output in an effort to satisfy the market's needs.

> The chief limiting factor on this strong demand is the ability of the industry to produce. For practical purposes it can be said that the industry is in fact producing at capacity levels and thus future near-term growth will be almost entirely dependent on increased production capacity.[1]

No one present doubted that more steelmaking capacity was needed. The problem was how to finance it. To examine the possibilities, one of the meeting's principal panel discussions dealt with methods for financing the expansion of the world's steel industry.

Steel demand continued at record levels during 1974 when expansion plans were crystallized. In October of that year, they were summed up at the IISI Munich meeting. The secretary general's keynote speech set the tone for the meeting. He said in part:

In the light of the very strong demand for steel which has continued unabated over the past two years and which still maintains a firm tone generally, I can find no basis for pessimism about the market outlook for the next 12 to 15 months.[2]

Every segment of the world steel industry represented at the meeting looked forward to a substantial increase in steelmaking capacity by 1985. Total new capacity planned for the non-Communist world amounted to 240 million tons, distributed as shown in table 2-1. Specific projects were undertaken or planned throughout the Western world to bring these projections to fulfillment.

Steel trade on a global basis prospered as some 170 million tons of raw-steel equivalent, or 24 percent of the world's output, moved into international trade. Every country was short of steel, and it was sought wherever it could be obtained. Prices for exports rose sharply—in some instances, by more than 300 percent of the early 1972 level. For example, plates from Western Europe to Third World countries increased in cost from $119 a ton FOB in January 1972 to $560 in August 1974. No price controls existed in the export market, and import restrictions were forgotten.

Despite the optimism, a note of caution was expressed by some of those in attendance at the Munich meeting. This caution was based on the energy crisis that was developing because of the fourfold increase in the price of oil. Some participants suggested that steel operations might fall by as much as 3 percent in 1975. However, few, if any, expressed alarm at this prospect. The prognosis was for an increase in steel output in 1975. The two-year steel boom had generated an optimism that was not easily dimmed.

The Collapse of the Boom

The prosperity of 1973 and 1974 ended abruptly in 1975, as the full impact of the fourfold oil-price increase, put into effect by the Organization of Petroleum Exporting Countries (OPEC) in late 1973, was felt in many quarters around the world. It dealt a severe blow to Japan and Western Europe, since they depended heavily on oil imports to meet their energy requirements, and the increased costs registered an immediate impact on their balances of payments.

In addition, the automotive industry that, in Western Europe, Japan, and the United States, is a major consumer of steel witnessed a sharp de-

Table 2–1

Total New Steelmaking Capacity Planned for the Non-Communist World: 1974

(millions of metric tons)

Countries	Tonnage
EEC	41.3
Other Western European	26.7
North American	28.5
Latin American	37.2
African	12.3
Middle East	23.8
Far East	67.9
Oceania	2.3
Total	240.0

Source: International Iron and Steel *Proceedings of the 8th Annual Conference of the International Iron and Steel Institute* (Brussels, 1974), p. 20.

cline in activity that continued into 1975. Worldwide automotive production fell from 39.2 million vehicles in 1973 to 35.0 million in 1974 and further declined in 1975 to 33.3 million vehicles.[3]

Construction and investment in capital goods, which are cyclical industries, also turned downward with a consequent adverse effect on steel production.

Another significant item that contributed to the sharp drop in steel output in 1975 was the unprecedented accumulation of steel inventories in the hands of consumers during the last half of 1974. In order to ensure themselves of adequate metal to meet their requirements, purchases were made well above current needs. Unfortunately, there was little, if any, appreciation of the extent to which inventories had grown. Further, steel customers placed duplicate and, in some instances, triplicate orders with both domestic and foreign steel companies, which filled the order books of the producers and gave them an exaggerated sense of steel demand.

In the non-Communist world, as steel production declined by a modest 6 percent in the first quarter of 1975, it soon became evident that steel was readily available, and since business was declining, many consumers began to live off their inventories. This combination of factors resulted in a drastic production drop for the year in the non-Communist world from the 1974 high of 494.0 million tons to 424.0 million, or about 14 percent.[4]

The United States and the EEC registered much greater drops of 20 percent and 19 percent respectively. Japanese output fell from 117.0 million to 102.0 million tons, or about 14.5 percent.[5] These developments sent shock waves through the steel industries of these areas as profits declined significantly and, in many instances, turned into substantial losses.

Trade restrictions that had been forgotten during the boom were again actively discussed.

Latin America and the Communist world swam against the tide as steel production in these two regions actually increased in 1975.

Unlike the boom that uniformly affected most of the steelmaking areas in the world, the collapse and subsequent seven-year steel depression have had varying impacts in different parts of the globe. In some areas, the reaction has been severe, accompanied by drastic losses and bankruptcies; in some it has been relatively mild, resulting in smaller profits; while in others, particularly in parts of the Third World, it has been scarcely felt.

Despite these differences, the problems previously discussed have arisen to plague the industry and will continue to do so. An analysis of the major areas throughout the world points up the relative degree of difficulty that is faced by each. The principal areas for analysis are Western Europe, North America, Japan, and the Third World. The USSR and the Eastern European countries are also discussed but in less detail. In addition, a number of other countries, such as Australia and South Africa, not included in these categories, are covered.

The analysis, which is made on a country-by-country basis, is not meant to be a complete treatment of any country's steel industry but, rather, an indication of its present condition, problems, and future plans.

Notes

1. Report of the secretary general, IISI Conference Report of Proceedings, October 1973, Johannesburg, South Africa (Brussels: IISI).

2. Report of the secretary general, IISI Conference Report of Proceedings, October 1974, Munich, Germany (Brussels: IISI).

3. Motor Vehicle Manufacturers Association, *Facts and Figures, 1981* (Detroit: Motor Vehicle Manufacturers Association, 1981), p. 7.

4. IISI, *Annual Statistical Reports 1981* (Brussels, 1981), p. 2.

5. Ibid.

3

Western Europe
Faces the Crisis

During 1975–1982, with the exception of a few short periods, the Western European steel industry has been mired in a depression, resulting in huge losses of hundreds of millions of dollars per year for many of its integrated producers. As a consequence, a number of companies have faced bankruptcy and have had to appeal to their governments for aid. This has been given and, in some instances, at least partial nationalization has resulted. Suggestions for a long-term solution to the problem include mergers; the elimination of obsolete high-cost capacity; price stabilization; and a freeze on expansion.

EEC

The current situation has its roots in the rapid expansion of steel capacity throughout the world from the late 1950s up to and including the 1973–1974 boom. This growth resulted in a fivefold increase in output for the EEC countries between 1947 and 1974. The boom of the early 1970s, which gave impetus to plans for further expansion, came to an abrupt end in 1975, as EEC steel production dropped by 19 percent, falling from the record high of 155.6 million tons in 1974 to 125.2 million.

In the first few months of 1975, a concerted attempt was made on the part of steel producers to maintain price levels in the hope that the recession would be short lived and that demand would strengthen. By the second quarter of 1975, with no improvement in sight and much idle capacity, the price structure began to weaken as discounts and concessions were offered to customers to keep the steel mills running and to maintain employment. The weakening price structure forced steel producers in April of that year to attempt to establish a floor under the price of reinforcing bars and to provide guidelines for several other steel products. In 1976, there was a brief revival, as production in the EEC increased by some 7 percent to 134.1 million tons. Prices, however, though somewhat improved, were still relatively weak.

In 1977, EEC production dropped back to the 1975 level, and this triggered a collapse in prices, which fell for some products by as much as 50 percent below the 1974 level.

Financial losses in 1977 were huge, amounting to a total of $4.0 billion for the steel producers of the EEC. This second setback in three years gave rise to an atmosphere of grim pessimism, in which a number of companies were seriously concerned about survival. Some maintained that it would be impossible for them to avoid bankruptcy if the situation did not materially improve, but they all agreed that something had to be done to stem the tide. The Simonnet plan, put in operation in 1976 that called for voluntary cutbacks, had failed, as had the attempt to stabilize prices in 1975.

The Davignon Plan

The situation worsened in 1977 so that by the fourth quarter, Viscount Davignon, the new commissioner for the EEC, secured the agreement of the nine member countries to adopt a comprehensive program, known as the Davignon plan, which had a twofold objective: (1) to raise steel prices within the community as soon as possible; and (2) over a longer period of time, to restructure the steel industry and to improve its competitive position.

To achieve the first objective, the program called for voluntary cuts in production and restrictions on low-priced imports, with quotas and prices assigned to outside countries. Production cuts based on the demand for steel were projected on a quarterly basis. The agreement on the reduction of output, which covered tonnages by product for each quarter on a plant basis, was not between the producers themselves but between the individual companies and the commissioner, Viscount Davignon. The production cuts, which varied in intensity on a product basis ranging as high as 25 percent, were related to a reference period to determine the basis from which a reduction would be made. For the most part, this period consisted of the output for the first six months of 1974 when all companies were presumed to be operating at full capacity. Some participants in the Davignon plan thought this unfair since their new capacity and improved facilities were not up to full production until 1975.

A concession in respect to price, which proved to be a weakness in the plan, allowed the producers to adjust list prices downward to match a quotation from either an outside country or another producer within the EEC. During 1977 and early 1978, these downward price adjustments were so widespread that published prices bore little relationship to those actually charged the customer.

The longer-run objective, restructuring the EEC steel industry, was to be achieved by the gradual reduction of obsolete capacity, coupled with the modernization of remaining facilities and substantial restrictions on expansion. In this respect, a number of expansion plans that had been drawn up during the prosperous days of the 1973–1974 boom were either scrapped or postponed. Some of these included:

The proposal by Usinor to increase the size of its Dunkirk plant in northern France from 8.0 million tons to 12.0 million tons;

The plan to double the size of Solmer's plant at Fos in southern France from 3.5 million to 7.0 million tons;

The plan that BSC had to increase its capacity to 34–36 million tons by 1980, with the construction, at various locations, of five large blast furnaces—one of these was built at Teesside but was a replacement;

The construction of a greenfield-site plant by Italsider in Reggio, Calabria.

The Davignon plan was implemented by the creation of Eurofer I, whose members were the integrated steel producers in the EEC. The plan functioned with varying degrees of success until the summer of 1980, when it virtually collapsed due principally to the weakening market. Further, some participants chose to produce more than their assigned quotas that they maintained were not adequate to permit anything but heavy deficits.

Article 58

As the losses mounted in the second half of 1980, considerable pressure was brought to bear on Viscount Davignon to declare a crisis in the steel industry and to apply Article 58 of the EEC treaty, which calls for mandatory reductions in production schedules supported by sanctions. Under the first Davignon plan, the reductions were voluntary. The suggestion to invoke the application of Article 58 was not unanimously received by the members of the EEC since a number of the companies, particularly in Germany, voiced strong opposition to it. A compromise was reached, calling for the application of Article 58 for a limited period up to June 30, 1981. The criterion for determining the tonnage base from which reductions were to be made consisted of any twelve-months' production (not necessarily consecutive) between July 1977 and June 1980. Given this latitude, the companies picked their best twelve months, and thus the reference points were quite high, so that even when a reduction of 35 percent for a particular product was called for, the resultant quota was still substantial. A number of companies did not reach their quotas and were able to exchange them with other companies. Exceptions were also granted from the tonnage allotments when slightly larger quotas were assigned to some firms. This caused some resentment among a number of the companies involved.

The application of Article 58 met with some success. However, dissatisfaction still existed about the quotas, and the prices set were substantially undercut. In the first quarter of 1981, actual prices were about 15 percent below list.

The New Plan

The successor to Article 58, which went into operation on July 1, 1981, was the result of negotiations among the EEC integrated steel producers with a new element added. The negotiations gave recognition to the organization established in March 1981 of the privately owned smaller mills and mini-mills, called the European Independent Steel Association (EISA), representing some sixty-one companies with an aggregate capacity of 20.0 million tons.

The organization, designed to replace Article 58, is known as Eurofer II. However, since it was not possible to negotiate quotas on a voluntary basis for flat-rolled products, the production limitations for sheet and strip and their derivatives, such as cold-rolled sheets, galvanized sheets, and tinplate, were placed under an extension of Article 58. All of the other products are under the Eurofer II agreement and have quotas established on a voluntary basis. Sanctions that consist of the imposition of substantial fines for violation of the agreement have been provided to ensure compliance with the voluntary quotas.

The extension of Article 58 was for one year, up to June 30, 1982, as was the Eurofer agreement, with wire rods placed under Eurofer II for three months. Prices for the new plan were established at a level below the previous Article 58 in the hope that they could be attained. The new arrangement called for prices to be monitored, which surveillance was to be extended to trading companies and service centers. The object was to eliminate discounts and to increase revenues significantly for most product categories. For example, the new plate prices brought $50 per ton above the actual price charged in the first half of 1981, while cold-rolled sheets increased by $60 per ton and wide-flange beams by about $65 per ton. There was relatively little difficulty in obtaining agreement by all concerned on the need for, as well as the amount of, price increases.

The point of contention that forced the extension of Article 58 for almost half of the steel output was the lack of agreement on the amount of flat-rolled steel production to be allocated to each company. The reference point from which production cuts were made for the products under Eurofer II was determined by a complex formula, involving aspects of past reference points as well as some additions. The Article 58 and Eurofer II agreements have since been reextended to June 30, 1983.

Price Stability

The plans that were put into operation to solve the Western European steel industry's problem dealt with both production quotas and prices. However, more attention was given to monitoring quotas and less to maintaining price levels.

Quotas are designed to reduce production, and theoretically, this should stabilize prices. However, the theory has not been reduced to practice. Unfortunately, even though production was low, the market was weak, and under pressure, companies yielded on price in order to maintain operations. As indicated, cuts were in the area of 15 percent below list prices. When one large company made a concession, this was usually enough to bring the prices of all others to the lower level.

The new plan established a strict monitoring system for prices, and if this is not followed, the plan, like the others, is doomed to failure. On July 1, 1981, with the installation of the new plan, prices had firmed up. They remained firm for three months, at which time, on October 1, there was a further increase of about 7 percent put into effect by the EEC countries, with the exception of West Germany that deferred the increase to November 1. These prices have been accepted by steel customers, and a new round of increases was announced for January 1, 1982.

For the continued success of the plan, it is incumbent upon the large, integrated producers throughout the EEC to maintain their prices, even in the face of temporary customer resistance. If, on the one hand, the prices can be maintained and the customers are convinced that their competitors will pay the same as they do, then orders will continue to be placed since none of them will be at a competitive disadvantage. If, on the other hand, a large producer grants concessions, it could well spell the doom of Eurofer II. The provisions of the new plan for monitoring prices should be helpful in achieving its objective.

Long-Range Rationalization

To achieve the long-run objective of the Davignon plan, a number of companies in the EEC have made plans and taken steps to restructure and revitalize their facilities. These steps include mergers, the elimination of obsolete capacity, and the modernization of existing facilities. If these plans are carried out, the steel industry in the EEC in 1983 or 1984 will be smaller than it is today and unquestionably more efficient. Further, with capacity reduced closer to the level of demand, the companies should function at a higher operating rate and produce profits.

Changes have been instituted and proposed in all of the member countries of the EEC, as the following analysis demonstrates. It is not meant to be a complete catalog of all the developments that are underway in the EEC steel industry; rather, it highlights the more-significant changes in facilities and corporate structure.

United Kingdom

The most dramatic reduction in capacity in the EEC has been that undertaken by the BSC, formed in 1967 by the nationalization of 91 percent of the country's steelmaking capacity. Fourteen companies operating twenty-two integrated steel plants, as well as other facilities, comprised the new entity that had a total raw-steel capacity of approximately 28 million tons. The private sector, those companies outside of BSC, consisted of electric-furnace operators having a combined capacity of 3 million tons for the production of both specialty and carbon steels.

During the steel boom of 1973-1974, BSC outlined an ambitious expansion program, contemplating a capacity of 34 to 36 million tons of raw steel in the first half of the 1980s. With the collapse of the steel boom and the subsequent prolonged depression, production and sales of BSC dropped sharply and losses mounted. In 1975, the first year of the depression, losses were £255 million. There was a slight improvement in 1976 as losses dropped to £95 million. However, losses increased in the following years, due partly to large severance payments, culminating in an all-time high of £668 million in 1980.[1] As a consequence, plans for expansion were abandoned, although significant capital expenditures were made during the entire period to replace facilities and improve operations. The most outstanding project was the construction at Teesside of a 10,000-ton-a-day blast furnace, complete with coke ovens, as well as sinter and pellet plants. The new facility, which went into operation in 1979, represented a capital investment of $800 million and was a replacement of three other blast furnaces at the nearby Cleveland Works.

In the years between 1974 and 1979, the size of BSC dwindled until its capacity reached 21 million tons. At that time, it announced that a further reduction in capacity would have to be made, bringing the new total of manned steel capacity to less than 15 million tons. This involved plant closures at Bilston, Corby, Shotton, Shelton, and Consett. The reduction in capacity was accompanied by a similar reduction in manpower as the number of employees dropped by 70,000 in 70 weeks. It should be stressed that the current 14.4 million tons are designated as manned capacity because some facilities, totaling between 2 and 3 million tons, could be reactivated within a relatively short time by increasing manpower.[2]

BSC is now a much leaner and more-efficient operation than it has been since its inception. Capital expenditures are restricted to those items necessary for improved operations, and the new chairman, Ian MacGregor, hopes to have the company back on a profitable basis by 1983. Part of this anticipated return to profitability is founded on the fact that the government, which held the debt of BSC on which interest must be paid, canceled a large portion of it, saving BSC £59 million annually in interest payments. BSC looks hopefully to the future with half the size of the original company but with modern, efficient facilities, greatly reduced manpower, and significant gains in productivity. Progress toward profitability was made in 1981 by cutting the previous year's loss in half.

The private sector of the U.K. steel industry grew from its original 3 million tons in 1967 to between 5.5 and 6 million tons in 1979. Several new companies were added, including Sheerness, with a capacity of 500,000 tons, and Alphasteel, with a melting capacity in excess of 1 million tons but a finishing capacity that can only absorb 800,000 tons.[3]

During the past year, approximately 1 million tons have been lost as two companies—the Patent Shaft Steel Works, with 400,000 tons, and Duport Steel Limited, with 600,000 tons—have been closed. Remaining reliable capacity in the private sector is approximately 4.5 million tons. This is made up of several large minimills as well as a host of small specialty-steel producers. Round Oaks Steel Works, with about 600,000 tons of raw-steel capacity, is 97 percent owned by BSC. However, according to law, unless it is 100 percent owned, it is still listed in the private sector.

West Germany

The steel industry in West Germany is by far the largest in Western Europe and ranks fourth in the world behind the USSR, the United States, and Japan. Its peak production was achieved in 1974 with an output of 53.2 million tons, which was roughly twice the production of France, the second-largest producer in the EEC. Since 1974, the capacity in West Germany has not been increased significantly so that actual reliable capacity in 1982 is, at best, 55 million tons.[4]

The industry consists of several major integrated companies including Thyssen, Krupp, Klockner, Salzgitter, Mannesmann, and Hoesch, a division of ESTEL. With the exception of Salzgitter, all of these companies are privately owned. A number of other companies that are electric-furnace producers make a substantial contribution to total tonnage. Principal among these is the Korf organization. In the Saar area, Rochling-Burbach has a capacity of more than 3 million tons, but this is part of the ARBED group that has headquarters in Luxembourg.

During the past five years, a considerable program of modernization has been undertaken by a number of German companies. Mergers and plant reorganizations have caused some restructuring of the industry. Thyssen, the largest company in the country, for example, reduced its capacity from

19 million tons to about 16 million through the elimination of obsolete open-hearth furnaces, some of which were replaced by electric furnaces. Notable instances of this were the steel shop at Oberhausen, which had seven open hearths that have been replaced by two electric furnaces, as well as that at Witten where four open hearths with approximately 500,000 tons of capacity were partially replaced by an electric furnace.

The Krupp steel organization has a total theoretical capacity of about 6 million tons and is one of the few companies that actually recorded more steel production in 1979 than it did in the boom year of 1974. This was due to a merger in the late 1970s with Stahlwerke Sudwestfalen (SSW), a producer of specialty and carbon steels with a capacity of approximately 1 million tons.

In 1977, Krupp improved its ironmaking capability substantially by constructing one large 11-meter blast furnace at its Rheinhausen plant to replace three small, relatively less-efficient units. In addition, open hearths were dismantled, as well as electric furnaces, some of which were replaced by modern units. The main steel-producing facility is a basic-oxygen shop built at Rheinhausen in the early 1970s with two 300-ton converters. A new 100-ton electric furnace has also been installed at SSW to replace smaller units.

Klockner, which was the fourth largest producer in 1980 with about 5 million tons of raw-steel capacity, has improved its facilities significantly in the last three years, principally by the replacement of three large open hearths by the KMS process, which involves a bottom-blown oxygen converter. This can use solid input up to 100 percent so it does not require blast-furnace support. At its optimum operating rate, it is capable of producing 1 million tons a year. Klockner also replaced an 80-inch-wide hot-strip mill, built in 1957, with a new 90-inch-wide mill in 1973.

Hoesch operates both open hearths and oxygen steelmaking facilities in Dortmund. The open-hearth furnaces, which represent a total capacity of 3 million tons, will be shut down and the oxygen steelmaking equipment will probably be relocated within the next few years. There are no plans to replace the open-hearth capacity. In addition, a plate mill and a structural mill in Dortmund will be closed.

Mannesmann is exclusively a producer of pipe and tubes. It ceased production of sheets in 1968, transferring that product to Thyssen in return for Thyssen's pipe operations. The steel is produced at its plant in Duisburg, where it operates five blast furnaces, four of them relatively small, and a steelmaking plant with two large basic-oxygen converters and three small ones. Total steel capacity is between 4.0 and 4.5 million tons. A small electric furnace is operated intermittently, and four open-hearth furnaces that had an annual rating of 720,000 tons have been closed down. Thus, virtually all of the steel is produced by the basic-oxygen process.

Salzgitter is the one large, integrated company in the West German industry that is government owned. It has two plants: one at Salzgitter, with seven blast furnaces and three basic-oxygen converters, that has a capacity of 3.9 million tons; and the other at Peine, where three blast furnaces provide hot metal for three 90-ton basic-oxygen converters, that has a capacity of less than 3.0 million tons. Total tonnage for the company is approximately 6.5 million tons. However, some of the blast-furnace operations will be closed down, and steelmaking capacity by 1983 will be reduced to 5.0 million tons. Efficiency will be increased through the installation of continuous-casting machines so that 80 percent of the steel produced by 1984 will be continuously cast.

Up until late 1981, the German steel companies, with the exception of Salzgitter, have refrained from applying for subsidies except those that were granted for research and development and capital investment. However, in late 1981 and early 1982, with the prospective merger between Krupp and Hoesch to form Ruhrstahl, as well as other structural changes, some of the German companies have applied for subsidies. This is particularly true of Hoesch and Krupp who plan to merge once Hoesch is divorced from Hoogovens and dissolves the ESTEL merger.

France

The French steel industry includes two large integrated companies, Usinor and Sacilor. In addition, a recently built integrated steel mill, Solmer, is owned jointly by Usinor and Sollac, which is controlled by Sacilor. The other steel company with a substantial capacity is Creusot-Loire.

In 1979, because of financial difficulties and huge losses, both Usinor and Sacilor were, in effect though not in name, nationalized as the government converted its debt holdings in these companies to equity. The 1981 election, which brought a Socialist government to power, resulted in the full nationalization of the two companies.

Both Usinor and Sacilor have carried on extensive modernization programs as well as reductions in capacity during the last three years. Usinor closed steelmaking operations at Denain and Longwy, reducing its overall capacity by more than 2 million tons. It has also improved operations at the Dunkirk Works so that 87 percent of that plant's steel is continuously cast, reducing manhours per ton of hot-rolled coils from 6 to 3.6. Present reliable capacity for the company is between 11 million and 12 million tons, not including the participation in Solmer.

Sacilor has undergone extensive modernization in its steelworks, eliminating Kaldo converters as well as basic Bessemer converters and concentrating its steel production at two plants—Gandrange, which has a 3-million-ton capacity, and Sollac, with the same capacity. All of the steel is made by the oxygen process.

In 1970, a new steel group, Solmer, was formed in France. It was participated in by Usinor (47.5 percent), Sollac (47.5 percent), and Thyssen (5 percent). A fully integrated, greenfield-site plant was planned at Fos on the Mediterranean that would ultimately have a raw-steel capacity of 7 million tons a year. The first phase consisted of two blast furnaces with a combined annual capacity of 3 million tons and a steelmaking plant with two oxygen converters with a 3.5-million-ton capacity. The principal finishing facility was a 90-inch-wide continuous hot-strip mill. In the second phase of construction, a third blast furnace and a third oxygen converter were to be added, bringing the raw-steel capacity to the intended 7 million tons.

The first phase was completed and in operation in 1974, and although plans were made to proceed with the second phase of construction, they were canceled as the steel depression of 1975 continued into subsequent years. At the present time, the plant is owned by Usinor and Sollac since Thyssen relinquished its 5 percent share. There are no current plans to resume construction on the second phase, so Solmer will continue as a producer of hot-rolled sheets with its present capacity.

The largest specialty-steel producer in France is Creusot-Loire, which operates six electric-furnace plants. One of these plants, Les Dunes near Dunkirk, has a 60-ton basic-oxygen converter for which iron is furnished by the Dunkirk plant of Usinor. The open-hearth furnaces have been replaced by electric furnaces at the Les Marais plant, where a continuous-casting machine will be installed in the near future. Another continuous caster is due for installation at Les Dunes. Most of the electric furnaces throughout the company are relatively new, having been installed in the past ten years.

The nationalization of the French steel industry, which took place in late 1981, has succeeded in replacing the chief executives in the two largest companies. The ultimate structure of the industry had not been determined as of mid-1982. However, there has been some discussion concerning the replacement of facilities withdrawn from operation with new capacity. Such a move could well have an effect on the supply-and-demand relationship in Western Europe.

Luxembourg

ARBED, located in Luxembourg, operates a number of plants including three fully integrated units. The company has grown in the last few years, acquiring either interests in or complete control of other Western European steel plants outside of Luxembourg. The ARBED Group now consists of its plants in Luxembourg, as well as large facilities in the Saar and Belgium where it controls Sidmar, the newest fully integrated plant in Belgium. Total capacity for the ARBED Group is in the area of 12 million tons, approximately 6 million in Luxembourg, 3 million in the Saar, and 3 million in Belgium.[5]

During the past five years, a substantial restructuring program has been in progress at most of ARBED's plants. In Luxembourg, a number of small blast furnaces have been eliminated, and iron production is concentrated in larger modern units, one of which is capable of producing 5,300 tons per day and was installed in 1979. Thus far, five small blast furnaces have been closed down permanently, and a number of others are due to follow. The same procedure will be followed by ARBED at Rochling-Burbach in the Saar in which ARBED has a 98 percent interest. Some ten small blast furnaces will be replaced by one or two large, modern units, to be completed in 1983.

Another company in the Saar, Neunkircher, which is 97 percent held by ARBED, has been reduced from an integrated operation to a finishing mill. Four blast furnaces and some small open hearths as well as three oxygen converters have been closed down, eliminating approximately 1 million tons of steelmaking capacity.

Steelmaking units in other ARBED plants have been updated as open hearths and converters were removed in favor of oxygen steelmaking facilities. This was particularly notable in the Saar, where several oxygen converters and open-hearth furnaces were replaced by a new modern oxygen steelmaking shop that went into operation in 1980. This shop, along with two continuous-casting machines, required an investment of over $200 million, part of which was a grant from the German government and part a guaranteed loan.

By 1983, ARBED will have 12 million tons of modern, efficient steelmaking equipment. Sidmar, ARBED's plant in Belgium built in 1966 on a greenfield site with two large blast furnaces and a 3-million-ton oxygen steelmaking shop, as well as hot- and cold-rolling facilities for wide sheets, is a logical candidate for expansion. However, the Belgian government will be slow in approving any new capacity in view of the crisis that has plagued the rest of its industry.

Italy

The Italian steel industry is composed of a number of electric-furnace companies and Italsider, a large government-owned company with three fully integrated plants, a blast-furnace plant for pig-iron production, and two electric-furnace operations. Total reliable capacity of Italsider is approximately 17 million tons, of which 16 million are accounted for by basic-oxygen steelmaking and approximately 1 million by its electric-furnace operations.[6]

As part of its complex, Italsider operates the Taranto plant in Southern Italy, which at present has a raw-steel capacity in excess of 11 million tons. It is a modern plant in every respect with oxygen steelmaking, continuous casting, two hot-strip mills, and two plate mills. The second hot-strip mill was installed in 1974 and the second plate mill in 1973. The current emphasis at Taranto is directed at improving efficiency rather than increasing size. To achieve this objective, Italsider has entered into a "know-how" agreement with Nippon Steel.

The remaining plants are not quite as modern as Taranto, but plans have been made to improve them. For example, at Bagnoli, a new 65-inch-wide hot-strip mill with a capacity of 1.0 million tons is under construction. This will be augmented by a continuous caster for slabs. At Genoa, six open-hearth furnaces have been replaced by oxygen steelmaking facilities within the last three years.

Italsider represents 60 percent of the steel industry in Italy when Terni and Dalmine, two of its subsidiaries with an aggregate capacity of 1.5 million tons, are considered.

The remainder of the industry is made up of electric-furnace operations, some large, like Fiat, with more than 1.0 million tons of capacity. This company produces a wide variety of products, including hot- and cold-rolled sheets, plates, and seamless tubes. Another large electric-furnace plant is operated by the Falck group, which also has over 1.0 million tons of capacity.

For the most part, the remaining companies are relatively small electric-furnace operations, which have been classified as minimills, ranging in capacity from 50,000 tons to as much as 300,000 tons. A number of these companies are located in the northern part of Italy and are known as Bresciani. They are facing problems at the present time due to the high cost of electric power and gas, as well as high interest rates. A number of them have maintained that they may have to face bankruptcy if conditions do not improve within a short time. Unlike the other steel industries in the EEC, the Italian industry has not reduced its capacity and has no plans to do so.

Belgium

The steel industry in Belgium encompasses several major companies and a few small specialty producers. The large integrated companies include Cockerill, Sambre, Sidmar, Boel, and Clabecq. Since 1975, the industry has had difficulties that have increased in intensity in the last two to three years so that the government, in 1978, stepped in and reorganized Cockerill and Sambre, the two largest companies. As a result, both companies are 50 percent state owned.

As the situation worsened in 1980, a study was undertaken by the Japanese recommending a merger of Cockerill and Sambre. The merger was to have resulted in a reduction in the combined capacity of the two companies from approximately 13 million tons to 8 million. These reductions have not taken place and the problem has assumed political proportions to the point where the government fell. Some solution will have to be found before the end of 1982 if the newly merged unit is to function smoothly. It now appears that the merged unit's total capacity will be 6.1 million tons, spread over four steelmaking facilities.

The most modern plant in Belgium, which started operation in 1967, is Sidmar, a member of the ARBED Group. It is a fully integrated 3-million-ton facility producing light, flat-rolled products. Two other companies, Boel and Clabecq, each have an approximate capacity of 1.5 million tons. Both produce plates and wire rods; in addition, Boel produces flat-rolled products and Clabecq produces structural sections.

A significant amount of replacement and modernization must be undertaken by the newly merged Cockerill-Sambre complex. Sidmar is modern and thus requires no major replacement, while Boel has recently installed new facilities including a continuous caster and a high-speed rod mill.

The Netherlands

The steel industry in the Netherlands is composed of two companies: a very large integrated steel company, Hoogovens, and a small electric-furnace minimill, NFK Staahl. The large integrated works is located at IJmuiden with a raw-steel capacity of 7.0 million tons. In 1972, Hoogovens merged with Hoesch in Germany to form ESTEL. However, negotiations are currently underway to dissolve this merger.

IJmuiden is one of the most modern steelworks in Europe, having been built since World War II. It is located on the seacoast, which is advantageous for importing iron ore as well as exporting finished products. The oxygen steelmaking process accounts for 100 percent of the plant's raw-steel output. The principal rolling facility is a modern 88-inch-wide hot-strip mill that is fed by slabs from a continuous casting unit.

Future plans call for the installation of a coil box on the hot-strip mill, more continuous casting, and a continuous-annealing line for sheets. An old hot-strip mill installed in 1953 will be closed down when the coil box is installed on the new mill.

The plant site provides ample room for expansion, should this be required in the future, but as of 1982, there are no expansion plans.

The NKF Staahl Company, a wholly owned subsidiary of Thyssen, operates four electric furnaces with a combined capacity of approximately 330,000 tons. The products are wire and wire rods.

Denmark

The steel industry in Denmark consists of one company, Det Danske Stal-valsevaerk (DDS). In 1973, the plant operated three open hearths with a capacity of approximately 400,000 tons. During the next two years, two electric furnaces were added, boosting total capacity to 1 million tons. Peak production of 863,000 tons was reached in 1978.[7] In the middle of 1980, the open hearths were closed down, and production is now exclusively from the electric furnaces with a rated capacity of 700,000 tons.

As a result of the elimination of the open hearths, as well as the installation of continuous-casting units for slabs, the plant is a modern installation whose principal product is plates. The furnace feed is 100 percent scrap, most of which is gathered within the country.

Ireland

The steel industry in Ireland consists of one company, Irish Steel Holdings. For a number of years it operated one electric furnace with an annual capacity of about 75,000 tons. In 1979 the decision was made to expand the steel-making capacity to 300,000 tons through the installation of an additional electric furnace. This was in place and functioning in 1981. The principal products are concrete reinforcing bars and merchant bars. Galvanized sheets are produced from purchased cold-rolled-sheet coils.

Raw Materials in the EEC

Western Europe was able to supply most of its iron-ore and coal requirements for steel production in the prewar and immediate postwar years. As steel output increased by more than fivefold in the postwar period, however, much more ore and coal were needed so that it was necessary to depend on imports to an ever-increasing extent.

Iron Ore

During the 1960s, as steel output rose, Western Europe's iron-ore reserves, with the exception of those in Sweden and France, were being rapidly depleted.

In 1950, there was little need to obtain iron ore from outside of Europe. The countries now in the EEC (excluding Greece) produced an aggregate 48.5 million tons of raw-steel, while their ore production amounted to 59 million tons with Sweden producing an additional 14 million tons. In 1960, steel production in the EEC totaled 97.5 million tons, double its 1950 level, while ore production amounted to 110 million tons.[8] This latter figure must be qualified since 66 million tons of it consisted of low-grade French ore with a 33 percent iron content. To provide for its needs in 1960, the EEC countries had to import 51 million tons of ore, 18 million of which came from Sweden. A total of about 33 million tons were supplied from a number of other sources including Africa and South America. In 1974, the year of peak steel production, when it totaled 155 million tons, the EEC countries produced some 66 million tons of ore, 54 million of which came from France. To meet its requirements, the industry throughout the EEC imported 140 million tons of ore, 27 million of which came from Sweden. In addition, large tonnages were brought in from Brazil, Australia, Liberia, and Canada. Within the EEC, France used much of its own ore and supplied 13.5 million tons to Belgium and Luxembourg, as well as 3.7 million to West Germany.

Outside of France, very little ore was produced in 1974, as production in the United Kingdom and West Germany dwindled to very small amounts. Table 3-1 traces the production of ore in the EEC, as well as in other Western European countries, for selected years from 1950 through 1979.

Most of the EEC countries have become heavily dependent on imports from overseas. West Germany, in 1974, consumed 62 million tons of ore, of which 58 million were imported. The United Kingdom, which at one time produced a substantial amount of its requirements, watched output fall to

Table 3-1
Production of Ore in the EEC for Selected Years: 1950-1979
(millions of metric tons)

Country	1950	1960	1970	1974	1979
France	30.0	65.9	57.4	54.4	32.0
United Kingdom	13.1	17.3	12.0	3.5	4.3
West Germany	11.9	18.9	6.8	5.6	1.7
Luxembourg	3.8	7.0	5.7	2.5	0.6
Belgium	0.05	0.2	—	—	—
Italy	0.5	1.2	1.2	0.8	0.2
Sweden	14.1	21.3	31.8	36.3	26.6
Spain	2.1	5.2	7.1	7.8	8.5
Norway	0.4	1.8	3.9	3.2	4.3
Austria	1.8	3.5	4.0	4.2	3.2

Source: American Iron and Steel Institute, *Annual Statistical Reports,* (Washington, D.C.).

to 3.5 million tons, while 20 million were imported. Italy with virtually no ore of its own imported 97 percent of its requirement in 1974. France, despite its domestic ore production, imported some 15.8 million tons in the same year, principally from Brazil, Mauritania, Sweden, and Liberia. The Netherlands depends on imports for all of its ore. In 1974, its principal sources were Sweden, Brazil, and Liberia. Belgium, likewise, depends on imports for virtually all of its requirements, which in 1974 came from Sweden, with Australia, Brazil, and Liberia supplying relatively small but significant amounts. Luxembourg was the only country in 1974 that obtained virtually all of its iron ore from within the EEC. It produced a significant tonnage domestically and imported the bulk of its needs from France. The Minette ore mines in France are located very close to the Luxembourg border, so the transportation costs are at an absolute minimum.

In the years since 1974, steel production has fallen significantly, as has the consumption of iron ore. In 1979, the output of 140 million tons of steel required 38 million tons of local ore production, as well as 137 million tons of imports from outside of the EEC. Of the 38 million tons, France produced 32 million, which represented a drastic decline from its peak output of 66 million tons reached in 1960. Sweden continued to be a major supplier, exporting some 20 million tons to the EEC. Imports from overseas sources, particularly Latin America, increased markedly, with the main supplier, Brazil, accounting for 30.5 million tons. Liberia added 15.6 million tons, while Australia, a relative newcomer to the European market, shipped 11.8 million tons. Another significant supplier was Canada with 17 million tons.

The decline in ore production in the EEC during the past twenty years evidently has increased dependence on outside sources that, with the exception of Sweden and small amounts from Eastern Europe, are all non-European. A number of steel companies anticipated this and made provision for it by purchasing or taking a participating interest in ore-mining ventures in other countries. Joint ventures have been established in Africa, Brazil, and Canada.

French ore, which constitutes an ever-declining portion of that used in the EEC, representing 32 million of 175 million tons in 1979, will continue to decline for a number of reasons:

The cost of production will increase as there are fewer tons over which to spread fixed costs.

The social costs of laying off workers and providing pensions for retired personnel are mounting.

The low-grade ores, about 31–33 percent iron content, which can be upgraded to 42 percent by sintering, are being replaced by higher-grade ores.

The principal outlets for French ore are France and Luxembourg and, to a very limited extent, Belgium and West Germany. During the past two years, ARBED in Luxembourg has cut back its consumption of this material and has imported higher-grade ores from Sweden and Brazil to the extent of 25 percent of its requirements.

It will be several years before Luxembourg ceases to use the French Minette ores since they are still available at a relatively low cost. However, should production continue to drop, as it has in the past five years from 54 million to 32 million tons, it is doubtful whether a production of less than 15 million tons could sustain the operation. Further, the shift to higher-grade ores by ARBED, a principal consumer, means much greater pig-iron-production potential, since it would be putting 60–65 percent iron-bearing materials into its blast furnaces instead of 42–44 percent material.

As of 1979, the principal suppliers of iron ore to the EEC were Sweden, Brazil, Liberia, South Africa, Canada, and Australia. There is no reason to suspect that any of these sources will disappear in the first half of the 1980s. However, their relative position could well change. Brazil is seeking an increased penetration of the EEC as is Australia. The other principal sources—Canada, West and South Africa, and Sweden—will continue their efforts to maintain their positions. To stay in the market, the Swedes have reduced the phosphorous in their ore by crushing, concentrating, and pelletizing.

With French production declining, it is quite evident that in the future the steel industry in the EEC will depend almost completely on outside sources for its iron ore. Table 3–2 gives the imports of iron ore to the EEC by country of origin in 1979.

Transportation of Iron Ore

The EEC's dependence on overseas sources for iron ore raises some questions about transportation costs and their effect on total costs of steel production. Shipping rates have been increasing during the past few years and, for some routes, have more than doubled since 1976. To a great extent these are determined by the size of the carrier that transports the ore from Brazil, Venezuela, Canada, South Africa, and Australia to the European seacoast ports.

One of the principal ports through which iron ore makes its entry into the EEC is Europort, near Rotterdam. This recently developed facility is capable of accepting 250,000-ton ships. Other deep-water ports to which iron ore is shipped include Dunkirk and Fos in France, IJmuiden in Holland, and a number of U.K. ports.

For plants that are located inland, ore unloaded at Europort must be reloaded and sent to its final destination. The inland plants that cannot be

Table 3-2
EEC Imports of Iron Ore, by Country of Origin: 1979
(thousands of metric tons)

Exporting Countries	Importing Countries					
	Belgium-Luxembourg	France	West Germany	Italy	Netherlands	United Kingdom
Algeria	837	112	—	285	—	50
Australia	783	1,679	6,132	1,196	409	1,567
Brazil	3,690	4,397	11,985	4,006	2,391	4,271
Canada	866	1,271	6,802	2,731	1,180	5,998
Chile	—	—	169	—	—	—
France	7,979	—	1,861	—	—	—
India	42	37	123	184	—	194
Liberia	1,862	1,822	7,592	3,155	913	855
Mauretania	1,144	2,868	769	1,791	—	828
Norway	140	309	2,450	—	—	—
Peru	—	91	52	—	—	—
South Africa	419	991	4,431	1,328	—	1,654
Sweden	6,923	3,000	6,830	—	1,711	1,045
Venezuela	1,208	463	1,714	1,606	—	1,281
Other	250	305	793	1,033	826	181
Total	26,143	17,345	51,703	17,315	7,430	17,924

Source: U.S. Department of the Interior, *Bureau of Mines Yearbook.*

served directly by large bulk-cargo carriers thus incur an additional penalty of several dollars per ton because of the transshipment. Consequently, the advantages of large bulk-cargo carriers, which have been helpful in the case of the Japanese, are not completely available to a significant portion of the European steel industry.

In the past five years, the cost of shipping a ton of ore from the various producing countries to Europort and other overseas locations has increased significantly. In 1976, the shipping rate from Brazil to Europort was $3 per ton using 250,000-ton ships. In 1981, this rose to between $6 and $7 per ton, or more than double. South African ore, which is shipped from Saldanha Bay to Europort, carries virtually the same charge as that of Brazil. The rate from Venezuela to Europort is considerably higher since the Venezuelans are limited to 60,000 ton ships. In 1976, it was $4 per ton and, by 1981, had increased to between $9 and $10 per ton, placing the Venezuelans at a distinct disadvantage. The shipment of ore from Australia in 250,000-ton carriers cost $5 per ton in 1976 and $9 per ton in 1981.[9]

The basic reason for the increase in freight rates is the dramatic increase in the price of fuel oil—from $80 to $180 per ton between 1974 and 1980. (As indicated elsewhere, the Japanese paid a higher price of $220 per ton in 1981.) High transportation costs are compensated for to some extent by the higher quality of imported ore as opposed to the ore that was mined in Europe. South American and Australian ore has a 63–65 percent iron content, which increased iron production in the European blast furnaces, as compared with the indigenous lower-grade beneficiated ore that, in some instances, had an iron content of 42–44 percent.

Iron-ore costs will increase in the future, particularly in view of the fact that the two main suppliers to the world, Brazil and Australia, are either losing money or just breaking even on their ore sales due to increased fuel costs for shipping, as well as port developments. The European steel industry will have to pay these increased costs because, although a number of countries are selling ore to Europe, the costs of production will continue to rise and force an increase in price despite the presence of considerable competition.

Coal

Western Europe, and the EEC in particular, imports a large amount of coking coal despite the fact that a substantial tonnage is produced by several of the member countries, with the greatest amount coming from West Germany.

In 1969, the West German steel companies, with the cooperation of the government, pooled their resources to form Ruhrkohle-Verkauf GMBH to produce coal to be converted into coke for the German steel industry. The new entity has the capacity to provide for most of Germany's needs and to export some coke and coal to countries within the EEC as well as outside of it. Ruhrkohle is one of the largest coal companies in the world with an annual output of 70 million tons. In 1974, it operated twenty-one coke-oven batteries, which by 1979 had been reduced to fourteen, thus eliminating some 5.0 million tons of capacity. In 1980, its capacity to produce coke was in the area of 17 million to 18 million tons, enough to supply most of the German steel industry's needs.[10]

Ruhrkohle is the principal, but not the only, coke producer in West Germany. Several batteries are operated by steel companies including Mannesmann, Salzgitter, and Thyssen, as well as a subsidiary of ARBED.

The West German supply of good coking coal was augmented by imports in 1980 and 1981. In the former year, some 1.3 million tons were brought in, of which more than 1 million came from the United States, with relatively small amounts from other EEC countries and the USSR. In 1981, approximately 1.5 million tons of coking coal were imported from the United States.

Italy is almost totally dependent on imports for its coking coal. In 1980, 16.5 million tons of all types of coal were imported, of which 6 million came from the United States, with South Africa supplying 3 million tons and Poland and West Germany supplying approximately 2.5 million tons each. In spite of the long distance, Australia contributed 1.1 million tons to the Italian coal supply. A substantial portion of total imports was used for the production of coke, with the remaining tonnage applied to other uses.[11]

France produces a significant tonnage of coal. In 1979, its output of hard coal amounted to 18.6 million tons, which was very close to the 1980 figure of 18.1 million. In spite of this production, imports amounted to 19.5 million tons in 1979. The largest amount, 8.4 million tons, came from South Africa, with Poland providing 4.5 million; the United States, 3.4 million; and Australia, 2.4 million. The South African tonnage almost doubled over a two-year period, 1977–1979.

Belgium was a significant producer of coal for many years. However, production of a number of its mines became uneconomical and they have been closed down. In recent years, hard-coal production, which includes both anthracite and bituminous, has been low, amounting to 6 million tons in 1980, so that it has been necessary to import coking coal. In 1979, imports amounted to approximately 4.5 million tons, a figure that was duplicated in 1980. In 1980, the United States supplied approximately 3 million tons, with West Germany providing approximately 800,000 tons. One of the Belgian companies, Sidmar, has a joint venture in a coal mine in the United States.

The United Kingdom, traditionally a coal producer, in the past decade has resorted to imports, particularly of high-quality, low-volatile coal. In 1980, the United Kingdom imported 7.3 million tons of coal, 3.6 million tons of which came from the United States, 2.8 million from Australia, and 446,000 from Poland. Of the total amount, 4.6 million was steam coal and 2.4 million was coking coal, of which 1.5 million came from the United States, 0.5 million from Australia, and 0.4 million from Poland.

Because of the large tonnages of imports, the National Coal Board in the United Kingdom sought to cut back on its coal stocks and to accelerate the closing of noneconomic mines. This program, proposed in March 1981, met with vigorous opposition and had to be modified, so that coal production is now subsidized.

BSC feels that it is necessary to import high-quality, low-volatile coal in order to produce coke that will perform in the large blast furnaces that have been operated in the last few years. This is particularly true of the 10,000-ton-a-day unit put in operation at Teesside in 1979. In the annual report for 1979–1980, the company states:

Action to eliminate BSC losses affected all aspects of costs. This action included a significant increase in the proportion of imported coking coal. Apart from quality considerations, the increase in imports was justified by the saving due to the disparity between the cost of home and overseas supplies of coking quality coals—a disparity which has considerably widened during the 1970s.[12]

It is precisely to correct this disparity that the subsidy was put into effect in 1981.

In addition to the coke production of BSC, the National Coal Board has a subsidiary, Coal Products Limited, that produces coke. The capacity of this company is approximately 3.2 million tons. It is committed to use domestic coal and has blended inferior coals with high-quality coals and successfully produced coke, some of which is sold to BSC and the remainder is shipped into the export market. This coke is made from a blend consisting of 45 percent high-quality coking coals, 35 percent low-quality coals, and 20 percent noncoking coals.

ARBED, Luxembourg's only integrated steel company, produces no coke at its plants located within that country. Its requirements are fulfilled by a subsidiary, EBV, located in Germany, in which ARBED holds 96 percent of the capital. This company has mines and coke ovens that supply the complete requirements of ARBED's blast furnaces in Luxembourg. In addition to EBV, ARBED has interests in coal mines in the United States where it formed Coal ARBED, which controls companies capable of more than 1 million tons of production.

In the Netherlands, the Hoogovens Division of ESTEL operates several batteries of coke ovens, the coal for which is supplied by imports, principally from the United States and West Germany. In 1980, the United States exported 2.6 million tons of metallurgical coal to Holland, a considerable increase above the 1.9 million exported in 1979. In 1980, the Netherlands imported 1.3 million tons of all types of coal from West Germany, some of which was used to make coke. In addition to coal imports, Hoogovens obtains between 500,000 and 700,000 tons of coke a year from West Germany.

The European steel industry evidently depends on imports of metallurgical coal to a very substantial degree. In 1980, the United States shipped 20.5 million tons of metallurgical coal into the EEC. Other countries that supply significant tonnages include Australia, Poland, and South Africa.

Attempts are being made to introduce inferior coking coals and limited amounts of so-called noncoking coals into the blends charged into the coke ovens. It remains to be seen whether these will be successful, as well as the extent to which they will be used. At any rate, it is quite clear that the EEC countries of Western Europe will continue to import sizeable tonnages of coking coal for the remainder of the present decade.

Scrap

In the EEC, four countries are heavily dependent on scrap as a feed material for steel production. Two of these, Italy and the United Kingdom, are major producers of electric-furnace steel. In 1979, a relatively good year for steel production in the EEC, Italy produced 53 percent of its steel, or 13 million tons, by the electric-furnace process, while in the United Kingdom, electrics accounted for 34 percent of production, or 7.4 million tons.[13] Denmark and Ireland depend completely on the electric-furnace, but their outputs are relatively small. The full capacity of the industry in Denmark is approximately 700,000 tons, while that of Ireland is 300,000 tons.

Scrap constitutes almost 100 percent of the charge for the electric furnaces in Italy and the United Kingdom. It is plentiful in the United Kingdom and, therefore, unnecessary to import this material. In fact, the United Kingdom is an exporter of scrap, sending as much as 1 million tons a year to Spain and smaller amounts to northern France, northern West Germany, and Denmark. Italy, however, has a substantial problem in providing enough scrap for its furnaces and is forced to import large quantities. It comes from Germany and France, as well as the United States. In 1979, Italy imported 1.2 million tons from the United States, as well as significant amounts from West Germany and France.[14] Because of the large tonnage of electric-furnace steel, Italy will be dependent on scrap imports into the foreseeable future.

The past and projected increases in continuous casting in the European steel industry, which now casts upwards of 45 percent of its production, have reduced and will continue to reduce the amount of revert or home scrap generated in the steelmaking process. Thus, there will be a greater dependence on obsolete metal that is gathered by the scrap dealers.

With respect to West Germany, France, Belgium, Luxembourg, and the Netherlands, there is no scrap problem since the percentage of electric-furnace steel is quite small, ranging from zero in Luxembourg to 3.3 percent in the Netherlands, and to 16 percent in Germany and 17 percent in France.

Technology

The steel industry in the EEC was built in great part after World War II. As previously indicated, production in 1947 was 29 million tons, increasing to 155 million in 1974, or better than fivefold. Further, almost half of the increase came after 1960. Consequently, the facilities are of relatively recent vintage, and the new greenfield-site plants such as Dunkirk in France and Taranto in Italy rank with the most modern in the world. By 1983, virtually all the prewar facilities will have been replaced, giving the EEC countries a modern steel industry. For example, during the past two years, a number of prewar blast furnaces have been shut down and several others will be eliminated in 1982.

Oxygen Steelmaking

The major technical breakthroughs in the steel industry since World War II have been applied extensively throughout the EEC. The most significant of these, the oxygen steelmaking process, was first brought to commercial production in Austria in 1952, and since that time it has increased to the point of dominating world steel production. Its growth in the EEC is evident from an analysis of steel output by processes given in table 3–3, which lists crude-steel production in the EEC by process for 1980. Table 3–4 indicates the growth of the oxygen steelmaking process as a percentage of total production for the countries in the EEC for selected years from 1968 through 1980. The year 1974 represents peak production for the countries in the EEC.

A number of modifications have been made in the oxygen steelmaking process. When first applied, it was referred to in Europe as the LD process and consisted of blowing oxygen through a lance placed above a bath of molten iron mixed with a percentage of scrap. Within a few years, a modification, referred to as the LDAC process, functioned in the same manner,

Table 3-3
EEC Crude-Steel Production, by Process: 1980

Country	Production (millions of metric tons)	Oxygen (%)	Electric (%)	Open Hearth (%)
Belgium	12.3	94.7	5.3	—
Denmark	0.7	—	76.2	23.8
West Germany	43.8	78.4	14.9	6.7
France	23.2	81.9	15.9	0.9
Ireland	0.03	—	100.0	—
Italy	26.5	45.3	53.0	1.7
Luxembourg	4.6	100.0	—	—
Netherlands	5.3	94.5	5.5	—
United Kingdom	11.3	59.3	40.6	—
Total	127.8	73.0	23.8	3.0

Source: International Iron and Steel Institute, *World Steel in Figures,* 1981
Note: France produced 1.3 percent of its steel by the Thomas process. This accounts for 0.2 percent of total EEC production.

Table 3-4
Growth of Oxygen Steelmaking in the EEC for Selected Years: 1968-1980
(percent of total steel output)

Country	1968	1970	1974	1979	1980
Belgium	38.7	53.6	79.6	95.7	94.7
West Germany	37.1	55.8	68.8	76.1	78.4
France	18.2	29.0	58.5	79.7	81.9
Italy	28.7	31.5	43.8	42.0	45.3
Luxembourg	33.2	37.6	65.1	100.0	100.0
Netherlands	62.7	77.8	92.2	94.5	94.5
United Kingdom	28.1	32.2	48.0	60.2	59.3

Source: International Iron and Steel Institute, *Steel Statistical Yearbook 1981* (Brussels 1981).
Note: Denmark and Ireland are omitted since they do not have oxygen steel production.

except that powdered lime, used to remove phosphorous from iron, was also introduced through the oxygen lance. The process is still in operation in Luxembourg, where a large percentage of high-phosphorous ore is used in the blast furnaces, and the phosphorous must be removed from the molten iron during the steelmaking process.

Further developments in oxygen steelmaking were the OBM and the LWS, in which the oxygen is blown through the bottom of the vessel, and if needed, lime can also be injected. Another process that employs both top and bottom blowing is the KMS, which injects powdered coal through the bottom of the vessel as well as lime, if needed. Claims for the bottom-blowing process include a higher yield of 1 to 2 percent because of less slag and an improved ability to produce lower-carbon steels. Another modification, described as the LDE process, injects inert gases, either nitrogen or argon, below the level of the bath to stir it and thus to improve production.

Tables 3-3 and 3-4 indicate the rapid increase in oxygen steelmaking in the EEC. Most of this capacity was installed as a replacement for the open hearth, which is rapidly becoming extinct. Table 3-3 indicates that only 3 percent of the steel produced in the EEC in 1980 came from the open-hearth process. This will be materially reduced by 1982, when Germany cuts its remaining open-hearth capacity substantially by closing a 3-million-ton shop operated by the Hoesch in Dortmund, and when Thyssen replaces 500,000 tons of open-hearth capacity at Witten with a large electric furnace.

By the mid-1980s, steel in the EEC certainly will be made entirely by the oxygen and electric-furnace processes. The division will most probably be 70-72 percent oxygen and 28-30 percent electric. It is also evident that the installation of oxygen steelmaking has virtually reached its limit so that very little, if any, expansion of this process is likely.

Continuous Casting

Another major technological breakthrough, which was perfected in Europe and adopted by steel producers in the EEC to a significant extent, is continuous casting. Although it has grown as a process, it is by no means as widely used as oxygen steelmaking. The major increase in continuous casting has come in the past five years.

In 1969, there was a negligible amount of continuous casting in the EEC except for West Germany where the process was applied to 7.3 percent of its steel output. In France, it amounted to 0.6 percent, in Italy 3.1 percent, and in the United Kingdom 1.8 percent, with nothing in Belgium, Luxembourg, or the Netherlands. By 1974, there had been a substantial growth, as table 3-5 indicates. Germany was the most progressive, moving to 19.4 percent, or 10.3 million tons: while Italy moved to 20.8 percent of its output, or 4.9 million tons. Since then, continuous casting has been widely accepted in the EEC to a point where in 1979 it was applied to 31 percent of steel production in 1981, this increased sharply to 45 percent.

Throughout the EEC, there is room for additional application of this postwar technological breakthrough, and judging from the thirty-one units

Table 3-5
EEC Continuously Cast Steel Output: 1969, 1974, 1975, 1979, and 1980
(percent of total)

Country	1969	1974	1975	1979	1980
Belgium	—	1.3	4.1	23.5	25.7
Denmark	—	—	13.1	58.8	73.3
West Germany	7.3	19.4	24.3	39.0	46.0
France	0.6	10.1	12.8	29.7	41.3
Italy	3.1	20.8	26.9	46.4	50.1
Luxembourg[a]	—	—	—	—	—
Netherlands	—	—	—	—	6.3
United Kingdom	1.8	5.5	8.4	16.9	27.1

Source: International Iron and Steel Institute, *Steel Statistical Yearbook 1981* (Brussels 1981).
[a]Luxembourg has recently installed continuous casting so that a significant percentage of its steel will be continuously cast by 1983.

currently on order for the member countries, it is virtually certain that by 1985 some 55 percent to 60 percent will be continuously cast. This will increase yield, improve quality, and bring about a significant rise in productivity.

Direct Reduction

The third major breakthrough in steel technology in the postwar period, direct reduction, has had limited application in Western Europe. The processes that dominate this development throughout the world are based on natural gas as a reducing agent, and this is relatively scarce in Western Europe. At present, there are five units in existence. Two of these, in the United Kingdom, have been completed but not yet put into production, due principally to the high price of gas that makes it uneconomical to operate them. The remaining three, in Germany, are currently functioning; however, one of these is scheduled for closure in 1982. In Italy, the solid-fuel Kinglor-Metor process for direct reduction, which has been installed in a few plants, produces very small tonnages and has a limited future.

    In general, the lack of cheap gas in Europe, particularly in the EEC, will tend to restrain the widespread use of direct reduction based on gas. A number of processes based on coal have been installed in several locations throughout the world, and further developments may permit their installation in the EEC sometime in the future. It should be pointed out, however, that the high percentage of oxygen steelmaking in operation in Germany,

France, Belgium, Luxembourg, and Holland keeps the demand for scrap or direct-reduced material, which is a supplement for scrap, at a low level. Consequently, the urgency to develop and introduce direct-reduction processes is not present.

In addition to these three technological breakthroughs, there are numerous other improvements, which, although very important, are not as basic. The use of computer technology is outstanding among these and is being applied to a great extent, as are other developments such as the AOD process that is used in the production of specialty steels, new blast-furnace tops, and improved refractories that allow a significant increase in blast-furnace temperatures.

A number of other technological improvements have been made to deal with pollution control. These in no way add to the quantity or quality of steel production but are necessary to comply with legal regulations concerning environmental controls in the various countries. They represent very large investments, particularly at the coke ovens, blast furnaces, and steel-making operations. A number of devices have been developed to control emissions both on the charging and pushing aspects of coke-oven batteries. These include pipeline charging and covering the entire coke-oven battery to restrict emissions when the coke is pushed. Another device includes a movable shield placed over each oven as it is pushed.

In respect to the blast furnaces, the cast house is the greatest pollution problem. Solutions include completely enclosing the cast house or covering the iron runners and drawing the pollutants out by a suction system. In respect to both electric and oxygen steelmaking furnaces, bag houses have been developed that serve to remove the emissions before they can be discharged into the surrounding atmosphere.

Ownership

The ownership of steel companies in the EEC has changed dramatically in the past fifteen years. In 1967, the U.K. government nationalized 91 percent of the steel industry and formed the BSC. This consisted of virtually all of the blast furnaces and integrated steelworks. Some fourteen companies operating twenty-two plants, rated at 28 million tons, were included in the nationalization. During the 1970s and early 1980s, the proportion has changed due to the contraction of BSC and an expansion of the private sector. As of 1981, BSC had some 14.4 million tons of raw-steel capacity, while the private sector had approximately 4.5 million tons, or 76 percent versus 24 percent.

In West Germany, the steel companies are privately owned with the exception of Salzgitter, which is a government-owned operation. In 1970, it merged with Peine and now has a total capacity of over 6 million tons, or slightly better than 11 percent of the German steel industry.

In France, the two large companies, Usinor and Sacilor, were privately owned until 1978, when a financial crisis required that the government step in and convert its debt to equity, giving the government 15 percent ownership in each company. However, through the government-owned banks, the public ownership extended to about 65 percent to 70 percent. For all practical purposes, the government owned both companies after 1978; however, it was not admitted officially that the companies were nationalized. At the start of 1978, Chatillon Neuves-Maisons was merged into Usinor, leaving very few companies outside of the government orbit. Creusot-Loire, a specialty-steel producer, is privately owned, as is Metallurgique de Normandie. Whether these will continue in private hands remains to be seen since the French Socialist government nationalized most of the steel industry in 1981, including Usinor and Sacilor.

The basic steel industry in Belgium consists of several integrated producers including Boel and Clabecq, which are 100 percent privately owned, and Sidmar, which is 62 percent owned by ARBED, with the remaining 38 percent scattered among other interests including the Belgian government.[15] The two largest companies, Cockerill and Sambre, have just been merged. Both of these were 50 percent government owned as a result of a capital restructuring in 1978. Between 1978 and 1981, they have received significant subsidies from the Belgian government, and although not technically nationalized, they are government dominated.

The largest steel producer in Italy, Italsider, is a government owned operation that represents 60 percent of the nation's steelmaking capacity. Other companies, including Fiat and the facilities operated by the Falck group, are privately owned as are a number of minimills.

ARBED is the only integrated steel company in Luxembourg and it is privately owned.

In the Netherlands, Hoogovens has its plant at IJmuiden, which is 27 percent owned by the government, 9 percent by the local town, with the remaining 64 percent in private hands.[16] The government does not run the company; its influence is limited to two directors on the board.

The steel company in Denmark, which has a relatively small capacity of 700,000 tons, was privately owned. In 1981, the government took a 30 percent share in the ownership as a result of injecting additional capital into the organization.[17]

Irish Steel Holdings, which has been expanded to 300,000 tons, is a government-owned operation.

A significant segment of the steel industry in the EEC has moved from the private sector to the public sector since 1967. Table 3-6 shows the shift from private to public ownership or domination. This has been substantial

Table 3-6
Steel Capacities Owned or Controlled by Government in the EEC:
1965 and 1981
(percent)

Country	1965	1981
Belgium	0	57
Denmark	0	30[a]
France	0	70[b]
Germany	10	11
Ireland	100	100
Italy	60	60
Luxembourg	0	0
Netherlands	0	36[a]
United Kingdom	8	76

Source: Compiled through company surveys in the EEC.
[a]In the case of Denmark and the Netherlands, one company is involved in each instance. The government ownership in those companies of 30 percent and 36 percent respectively does not mean government control but only representation on the board of directors.
[b]This number could change in 1982 if the French Socialist government decides to nationalize more of the steel industry.

and the future of the European industry seems to indicate more government intervention.

EEC Trade

From the mid 1960s through 1974, exports from the EEC countries increased substantially. The total for the nine countries rose from 33 million to 62.7 million tons between 1966 and 1974, with a steadily upward progression. Imports by the nine countries followed the same steady increase, although the tonnages were much smaller. Table 3-7 gives imports and exports for the EEC countries from 1966 through 1974.

A very large percentage of both exports and imports were exchanged among the EEC countries. Because of the nature of the EEC, steel flows across international borders almost without restriction, so that any one country looks at the entire group as its market. There are, of course, geographical problems in terms of freight when long rail hauls are involved.

Exports outside of the EEC in a number of years represented about 50 percent of total exports. For example, in 1979 the EEC countries exported 62.3 million tons of which 30.8 million were within the EEC boundaries and 31.5 million were directed to outside countries.[18] Areas that presented sig-

Table 3-7
Steel Imports and Exports for EEC Countries: 1966-1974
(millions of metric tons)

Year	Imports	Exports
1966	19.1	33.0
1967	21.2	36.6
1968	24.2	40.5
1969	29.2	41.4
1970	31.1	41.4
1971	29.8	45.4
1972	34.6	49.7
1973	35.4	54.5
1974	35.1	62.8

Source: International Iron and Steel Institute, *Statistical Reports* (Brussels, various years).

nificant markets to the EEC countries were other Western European countries, the USSR, Eastern Europe, North America, and the Middle East.

The bulk of the imports to the EEC, which amounted to approximately 10 million tons in 1979, came from other Western European countries, accounting for 5.7 million tons, or about one half. The principal suppliers were Austria, Sweden, and Spain, accounting for a total of 3.5 million tons. The Eastern European countries and the USSR accounted for 3 million tons, most of which came from Romania, Poland, and Czechoslovakia, with the USSR contributing very little. Countries outside Europe shipped small amounts into the EEC, including Japan, that provided only 800,000 tons in 1979.

Since 1975, when the current depression started in Europe, the EEC has protected its borders against imports with a series of restrictions, including tonnage quotas and established import prices. These were set up against the Japanese as early as 1972 but were allowed to lapse in the boom years of 1973 and 1974 only to be reinstated after the boom collapsed in 1975.

The U.S. market is a prime target of EEC producers. Since the mid 1960s, shipments have been substantial, ranging from a low of 2.9 million metric tons in 1976 to a high of 6.8 million metric tons in 1978.[19]

Most of the countries in the EEC have built steelmaking capacity well beyond their domestic needs and must export in order to operate profitably. To maintain these exports in years of sluggish world demand, a number of European producers cut prices sharply. This produced a reaction among importing countries where devices such as the trigger-price mechanism were instituted in order to increase the price of imports thereby reducing tonnage. Further, many of the traditional EEC export markets in the Third World have shrunk as these countries have either built or expanded their own steel industries.

Another limiting factor, particularly in the early 1970s, was the Japanese movement into the EEC peripheral markets with large tonnages. An example of such a market is Spain, where Japanese exports increased from 77,000 tons in 1971 to 887,000 tons in 1976.[20] Since that time, due to the rapid growth of the Spanish industry, exports have diminished; however, for the two-year period of 1975 to 1976, the penetration of the Spanish market was indeed large. Another example was Greece, to which Japan shipped over 500,000 tons of steel products in 1974. The Japanese also penetrated the Middle East markets of Iran and Saudi Arabia.

To advance the cause of orderly marketing and maintain export tonnage, discussions were held between some EEC producers and Japanese to control export tonnage to each other's peripheral markets.

For the rest of the 1980s, the EEC will continue to export significant tonnages of steel products with the United States as its principal objective.

The Basic EEC Problem and Future Outlook

The fundamental issue dividing the Western European steel industry today is that of the subsidized and government-owned companies versus the privately owned. The latter group is represented principally by most of the German steel producers, as well as ARBED, Hoogovens, and relatively small segments of the British and Italian industries. The principal government-owned or subsidized companies are the major producers in the United Kingdom, France, Belgium, and Italy.

The private companies have argued vigorously that they have been placed in an unfavorable competitive position due to subsidies granted to steel companies by the British, Belgian, Italian, and French governments, particularly when these subsidies are used to meet operating losses. It is maintained that, as long as a company can have its losses underwritten, it can afford to sell at low prices, whereas if it had to make ends meet, it would have to raise its prices. This condition has been blamed for the chaotic price practices of the European industry between 1975 and 1981.

It has been estimated that in the years from 1975 through 1983, a total of $30 billion will have been received in subsidies by state-owned or -subsidized companies. The nonsubsidized companies maintain that a sizeable share of this amount has kept high-cost, obsolete capacity in operation when, in fact, subsidies were designed to eliminate such capacity.

A number of meetings have been held by the ministers of the EEC to work out a program that would put an end to the subsidies. One of the most

important took place on June 24, 1981, when it was decided that all subsidies would end in 1985 and that subsidies for emergency purposes, covering operating losses, would end in 1982.[21] However, some conditions would
allow for exceptions, provided the approval of the EEC Commissioner can
be obtained. In the dispute over subsidized operations, the Germans have
threatened to impose an import duty on steel coming from other European
countries that have been subsidized in terms of operating costs.

Most of the Western European companies are applying for government
assistance to restructure their operations and improve their facilities. This is
taking place throughout the EEC and includes Germany, which has recently
been negotiating for a $541 million subsidy to aid in the restructuring of the
country's steel industry.

A struggle is in progress in Italy where small electric-furnace producers
maintain that they have been denied the necessary government help that
many of them need to continue in business. Without it, they complain
bitterly that they will have to liquidate or place themselves in receivership.
The organization representing the Italian minimills cited the high cost of
electricity and natural gas and bank interest rates as particularly pressing
problems. The president of the minimill association stated:

> We are proud of our modern and efficient plants but to operate them we
> must import half of the scrap and pay for energy up to 40 percent higher
> costs than at German and French mills.[22]

The Italian government, through the Interministerial Committee for
Industrial Policy, has recently approved the sum of $7.0 billion to be ex-
pended over the next five years to restructure and strengthen the steel industry. Funds are to be granted only to those companies that can show some
hope of a profit. Further, concerns receiving funds after mid 1983 will only
be granted them after they have shown successful industrial reorganization.

The controversy between the subsidized and nonsubsidized companies
will no doubt continue as long as the steel industry in Western Europe
remains in the doldrums. The private companies maintain that they have
had to keep their plants modern and efficient in order to survive, while their
subsidized counterparts are using subsidies to cover operating losses from
plants and equipment that are obsolete. Further, they stated that withdraw-
ing the subsidies will force all companies to charge higher prices to cover
costs. The issue has divided the Western European steel industry, and
attempts to solve the problem have not yet met with success. The new timetable worked out by the EEC ministers to end subsidies may well provide
the answer.

The short-run future of the steel industry in the EEC and indeed
throughout Western Europe is not too promising. The majority of the integrated steel companies have incurred and are continuing to incur huge losses
as a result of which it seems virtually certain that by 1985 almost all of the

integrated companies will be assisted by their respective governments. There will be some mergers. The Krupp-Hoesch union is probable and there are a number of other possibilities.

Total raw-steelmaking capacity in the EEC will probably decline to 150–155 million tons as obsolete blast furnaces, the remaining open hearths, and some uneconomical oxygen converters are shut down. Rolling mills that are inefficient will be replaced, and continuous casting will increase. The cost of these capital improvements is very high and will be shared by the respective governments. As a consequence of these developments, by 1985, the EEC, as well as other countries such as Spain, Austria, and Sweden, will have smaller but more efficient steel industries.

In terms of raw materials, particularly iron ore, the EEC is rapidly moving to the same position the Japanese hold. By 1985, more than 80 percent of its iron ore will come from outside of Europe. Sweden will continue to provide a relatively small amount, but there will be an increasing dependence on Brazil, Australia, Africa, Canada, and Venezuela.

Trade with the United States, particularly in years when general demand is sluggish, will be regulated by a negotiated agreement that will necessarily limit exports to that market. However, if world steel demand increases by even a modest 1.5 percent during the next few years, the European industry will, by the mid 1980s, operate at a high rate of capacity and, with its more efficient equipment, hopefully, can turn its losses into profits. For the next year or two, however, prospects are not particularly bright.

The steel industry in the EEC has witnessed a significant decline in employment in recent years. Between 1974, the year of peak production, and December of 1981, total employment in the steel industry for the EEC, exclusive of Greece, fell from 795,000 to 550,800. Most of the countries participated in this decline. In Germany, employment fell from 232,000 to 186,700; in France, it declined from 157,800 to 97,500; in the United Kingdom, it fell from 194,300 to 88,300; in Belgium-Luxembourg, the decline was from 112,300 to 75,500. The only exception was Italy, where employment actually increased from 95,700 to 97,500.

During the next few years, at least up to 1985, if the countries in the EEC carry out the reductions in steel capacity that they have projected, particularly in Belgium and Germany, employment will decline further. Thus, the EEC steel industry will not present an employment opportunity in the decade ahead.

Other Western European Countries

The countries in Western Europe outside of the EEC accounted for approximately 34 million tons of steel production in 1980. This figure represents a substantial growth during the past fifteen years, increasing from 15.7 mil-

Table 3-8
Crude-Steel Production of Non-EEC Western European Countries:
1965 and 1980
(millions of metric tons)

Country	1965	1980
Austria	3.2	4.6
Finland	0.4	2.5
Greece[a]	0.2	1.2
Norway	0.7	0.9
Portugal	0.3	0.7
Sweden	4.7	4.2
Spain	3.5	12.7
Switzerland	0.3	0.9
Turkey	0.6	2.5
Yugoslavia	1.8	3.6

Source: International Iron and Steel Institute, *Statistical Reports* (Brussels).
[a]Greece joined the EEC on January 1, 1981.

lion tons in 1965 to the present level that is more than double. The ten countries included in this group, along with their 1965 and 1980 steel production, are listed in table 3-8.

The three principal producers—Spain, Austria, and Sweden—accounted for 63 percent of total output producing steel by the basic-oxygen, open-hearth, and electric-furnace methods. The amount of open-hearth production is extremely small in all three cases, and it could well be eliminated within the next two to three years. The next three largest producers—Yugoslavia, Finland, and Turkey—had 25 percent of the total. In terms of steel-making technology, two of these countries, Turkey and Yugoslavia, still have a heavy dependence on the open-hearth process, while Finland produces a very high proportion of its steel by the oxygen process. Table 3-9 indicates the percentage contribution of each process to the various country's total steel output. In regard to continuous casting, most of the countries cast a significant percentage of their steel, with Turkey and Norway the two exceptions, as table 3-10 indicates.

In terms of raw-material requirements, with the exception of Sweden and Norway, both of which are exporters of ore, the other countries have to depend on imports for a substantial portion of their ore needs. Coking coal must also be imported, with significant amounts obtained from the United States. In 1979, the United States provided Yugoslavia with 1.1 million tons, while Spain took 1.4 million; Portugal, 0.3 million; and Norway, 0.2 million tons.

The non-EEC Western European countries, considering their size, export significant steel tonnages. In 1980, the total for these countries was

Table 3-9
Crude-Steel Production, by Process, for Non-EEC Western European Countries: 1980
(percent of total)

Country	Open Hearth	Oxygen	Electric
Austria	3.3	83.9	12.8
Finland	—	87.3	12.7
Norway	—	53.8	46.2
Portugal[a]	—	57.2	39.9
Spain	5.0	45.8	49.2
Sweden	4.9	50.7	44.4
Turkey	51.8	22.6	25.6
Yugoslavia	40.1	31.6	28.3

Source: International Iron and Steel Institute, *World Steel in Figures,* 1981.
[a]Portugal produces 2.9 percent of its steel, or approximately 20,000 tons, by another process.

Table 3-10
Steel Continuously Cast, by Non-EEC Western European Countries: 1980
(percent of total production)

Country	Production
Austria	51.2
Finland	90.2
Norway	12.9
Portugal	43.1
Spain	36.7
Sweden	49.0
Turkey	—
Yugoslavia	36.6

Source: International Iron and Steel Institute, *World Steel in Figures 1981* (Brussels: 1981).

12.2 million tons, a large increase over the 5.2 million tons of steel products exported in 1971. Imports in 1980, striking a balance for the group, were 13 million tons. Individual countries differ substantially, however. For example, Austria exported 2.4 million tons, while importing 900,000 tons; Switzerland exported 566,000 tons, while importing 1.9 million tons.[23] In most instances, the EEC was the main recipient of these tonnages. More than 80 percent of the tonnage produced in the Western European non-EEC countries comes from government-owned mills.

In terms of growth, a few countries such as Turkey and Yugoslavia have plans for expansion. However, most of them are content, at present, to remain with current capacity that is operating at much less than its full potential.

Austria

In relation to its size and population, Austria had what could be called a significant steel industry before World War II. The high point in its production was 637,000 tons in 1928, after which it suffered severely in the depression, recovering to 636,000 tons in 1937.[24] From that time until the end of World War II, its tonnage is included in the German figures.

A fully integrated plant, with some six blast furnaces, was built at Linz in the years immediately before and during the first part of World War II. Its corporate structure was reorganized in 1945 to form Voest, the largest steel company in Austria.

Austrian steel production grew rapidly after World War II, reaching 1 million tons in 1951. By 1960, it was 3.1 million tons, after which time it continued to increase to a high point of 4.7 million tons in 1974. Along with the rest of Western Europe and the industrialized world, output dropped in 1975 to 4.1 million tons. However, by 1979, it had recovered to surpass previous production levels as it reached 4.9 million tons.[25]

The steel industry in Austria, as it is constituted today, consists primarily of one company, Voest-Alpine, the result of a merger in 1973 between Voest, which was organized in 1945, and Alpine, a company whose origins date back to 1881. In recent years, one of the leading specialty-steelmakers, Schoeller-Bleckmann, was absorbed into the Voest-Alpine combine. The capacity of the combine was in the area of 4 million tons as of 1981. In addition to Voest-Alpine, several small electric-furnace plants are in operation.

In the years immediately after World War II, the Voest plant at Linz operated five blast furnaces and an open-hearth steelmaking shop, as well as some small electric furnaces. In 1952, this plant made history as the first in the world to install the basic-oxygen process, when three thirty-ton vessels were constructed. The Donawitz plant of Alpine was the second to install this process, with two thirty-ton vessels, in 1953.

In 1977, a new 5,500-ton-per-day blast furnace with an 11-meter hearth was constructed at Linz. This replaced three of the smaller units that had been in operation since the early 1940s. During the past two decades, additional oxygen converters have been installed at Linz. Until recently, three shops were operating, including the original with three 30-ton converters, a second shop with three 55-ton converters, and a third shop with two 135-ton converters. The first shop has been closed down, as well as the open-hearth furnaces, so that the plant's steel is now made in the second and third basic-oxygen shops. Future plans call for the replacement of number 2 shop, with its three 55-ton vessels, by one large 230-ton converter. This will be in operation in 1984; however, since it is a replacement, there will be no additional steel produced. At Donawitz, the two smaller vessels, originally installed in 1953, have been replaced by two 60-ton vessels. In terms of finishing facilities, Linz operates plate and sheet-production units, while Donawitz concentrates on the long products.[26]

Voest-Alpine is a company with a wide range of industrial capabilities, not only in steel but also in engineering and construction. It builds steel mills, turbines, nuclear vessels, and hydro-electric power stations.

Most of the raw materials for steel production come from outside the country, with imports accounting for 80 percent of the ore requirements and virtually all of the coking-coal needs. More than half of the ore comes from Brazil, with additional amounts from the USSR, Canada, and Liberia, while some 20 percent is supplied by upgrading indigenous ores. Coking coal is imported from West Germany, Poland, the USSR, and Czechoslovakia.

Austria exports a considerable amount of its steel production. In 1974, for example, exports were 1.7 million tons; by 1979, they had risen to 2.4 million. Imports, however, were considerably less than 1 million tons in both years. The principal recipient of Austrian steel is the EEC, to which 1.4 million tons were shipped in 1979.[27]

In terms of technological development, Austria has unquestionably made a most outstanding contribution to the world steel industry with the oxygen-steelmaking process. Since its institution in 1952, it has become the dominant steelmaking process throughout the world and, by 1985, will probably produce about 65 percent of the world's steel. In Austria, it accounts for 84 percent of the nation's output.

No additional capacity is planned for the foreseeable future for Voest-Alpine, which is 100 percent state-owned, so that by 1985, Austrian production will most probably not exceed 5 million tons.

Spain

Spain is the leading steel producer in Western Europe outside of the EEC. This position was achieved in the late 1960s and has been strengthened since then, as table 3–11 indicates. The 1980 figure of 12.6 million tons is partic-

Table 3–11
Spanish Raw-Steel Production for Selected Years: 1946–1980
(millions of metric tons)

Year	Tonnage
1946	0.7
1955	1.2
1965	3.5
1970	7.4
1974	11.5
1980	12.6

Source: International Iron and Steel Institute, *Statistical Reports* (Brussels).

ularly significant, since it indicates that Spain surpassed its 1974 peak output by a considerable amount, while all of the EEC countries, with the exception of Italy, failed to reach their 1974 peaks in subsequent years.

The industry has a raw-steel capacity in 1982 of 15 million to 16 million tons, about half of which is accounted for by three integrated steelmaking complexes and the remainder by a number of electric-furnace producers. The largest of the integrated companies is Ensidesa, a government-owned company, that operates two plants, one at Aviles, with a capacity in excess of 3 million tons, and another at Verina, with a 2-million-ton-plus capacity. The Verina plant was absorbed by Ensidesa in the early 1970s when it took over Uninsa, another government-owned enterprise.

Current plans for Ensidesa call for a new, large oxygen steelmaking shop at Aviles, which when installed will eliminate the open-hearth furnaces and will provide some additional tonnage over its 1981 capacity. More continuous-casting units will be installed at this plant, and improvements will be made on the hot-strip mill.

The second largest integrated company is Altos Hornos de Vizcaya. Its current capacity of 1.5 million tons will be expanded to 2 million tons when the three 70-ton oxygen converters now operating are replaced by three 100-ton converters. Continuous casting will also be added.[28]

The third integrated company, Altos Hornos del Mediterraneo, had a theoretical capacity to produce some 900,000 tons of raw steel. However, effective capacity at the present time is 600,000 tons. Long-range plans call for this capacity to be phased out, with steel to be provided by Ensidesa as it expands its operation. Altos Hornos del Mediterraneo has installed a thoroughly modern cold-reduction mill, which is provided with hot-rolled bands from sources outside the company.

The Spanish steel industry has employed the latest technology in terms of oxygen steelmaking and continuous casting. Some of the smaller electric-furnace companies are also installing facilities for the direct reduction of iron ore. During the next five years, the modest expansion plans and the phasing out of some facilities suggest a regrouping rather than an increased steelmaking potential.

During the past decade, Spain has increased its exports considerably, corresponding to its increase in steel production. In 1970, exports were 267,000 tons. They jumped to 918,000 tons in 1971, and since then they have continued to increase. By 1977, they were 2.7 million tons; in 1979, they rose to 4.2 million tons. Imports were 1.8 million tons in 1970, declining considerably in the following years, except for 1976 when they increased abruptly to 2.6 million tons. By 1979, they had dropped to 1.1 million tons.[29]

Spain will unquestionably continue to be a sizeable net exporter of steel during the remainder of the decade, although the tonnage may drop below the 4.2-million-ton high. As domestic needs increase, more of the steel produced in the country will be allocated to satisfy them, and this should reduce exports, particularly since there are no significant plans to increase overall steel capacity during the next five years.

Sweden

In May 1978, a plan to restructure a large segment of the Swedish steel industry was put into effect as the three principal steel producers, Stora-Kopparberg, Granges, and Norrbottens, were combined to form Svenskt Stål Aktiebolag (SSAB). The new corporation is dominated by the state with 50 percent ownership, while Stora-Kopparberg and Granges each hold 25 percent. It is expected, however, that Stora-Kopparberg will soon sell its 25-percent share to the state. The new combine operates steelworks at Domnarvet, formerly owned by Stora-Kopparberg; a fully integrated steel mill at Luleå, formerly owned by Norrbottens; and a fully integrated plant at Oxelösund, formerly owned by Granges. In addition, it has iron mines in central Sweden.

In order to streamline the operation, it was decided to concentrate pig-iron operations and oxygen steelmaking at Luleå and Oxelösund and to restrict Domnarvet to electric-furnace operations. Also, finished products were distributed among the three divisions.

The purpose of forming SSAB as stated was:

[t]o bring together the three largest steel mills in Sweden and their associated operations into a single company so that the necessary structural changes can be implemented to create a competitive and profitable steel company with a capacity to survive in the future.[30]

Financial results of the Swedish steel companies in the postboom years of 1976 and 1977 indicated the need for restructuring to preserve the industry. The Swedes were quite concerned about the future viability of their major steel companies, as well as their ore production, because of developments in the steel industry in other sections of the world. With this in mind, the new company was directed to aim at a market in the Nordic countries and hope for some exports. In setting the objectives for the new combine, it was stated that:

The plants must be concentrated, specialized, and given the right material and product flows in order to achieve economies of scale. Furthermore, efficiency and product quality must equal those of the best Western European mills. Facilities shall be sized so that high capacity utilization can be achieved in separate stages. Investments in profitable and high-poten-

tial product areas at higher levels of production must therefore be given
priority over investments at lower production levels. As a consequence, the
concentration of investment was to be on finishing facilities.[31]

Losses were anticipated during the formative years of the new enterprise, and a reconstruction loan was set up to cover them. This amounted to
1.8 billion Swedish kronor, or approximately $350 million. It could be
drawn down between 1978 and 1982. Thus far, losses have been relatively
heavy, amounting to a total of some 1.7 billion kronor from 1978 through
1980.

The reconstruction loan was based on the condition "that SSAB carry
out structural changes appropriate to the creation of a competitive and
profitable steel undertaking."[32] It is interest free for three years, after which
it carries 9.5 percent interest. Further, "the loan may be written off to the
extent of 50 percent of the sum disbursed at the end of the second year after
the calendar year in which the disbursement took place. Borrowed amounts
that have not been repaid by the end of the ninth calendar year after a
calendar year in which they were disbursed may be written off in their
entirety."[33]

In addition to the reconstruction loan, the state provided a structural
loan of 1.3 million kronor, which could be drawn during the years 1978 to
1980. "It runs without interest for three years after which it carries interest
at 5 percent p.a. No repayments are required for the first five years, after
which the loan must be amortised over a period of twenty years."[34]

SSAB plans to install continuous-casting equipment at Luleå and Oxelösund, while improving the hot- and cold-strip mills at Domnarvet. No other
sizable investments are planned. Thus, the ambitious plan, conceived in the
early 1970s for Norrbottens, that would raise the capacity of the plant to 4
million tons has been abandoned.

The production capacity for the new company is approximately 3.6 million tons of crude steel, with 1 million at Domnarvet, 1.1 million at Oxelösund, and 1.5 million at Luleå. Output since its formation in 1978 has been
a consistent 2.2 million tons annually. Total output for the country in the
same period was an average of 4.4 million tons.[35]

In addition to SSAB, the Swedish steel industry consists of a number of
smaller companies capable of producing 2.2 million tons of steel; of this,
approximately 500,000 tons are of carbon steel and 1.7 million tons of specialty steel. In the specialty-steel area, there are a number of producers, the
largest of which are SKF Stål and Sandvik AB.

During the past fifteen years, there has been no growth in Swedish steel
production. In 1965, it was 4.7 million tons. After this time, it increased
slightly, hovering between 5 million and 5.5 million tons until 1974 when it
reached a high of 6 million tons. Since that time, it has declined and, in
1980, stood at 4.2 million tons. The Swedish steel industry is content to

maintain this level of operations, concentrating on improving efficiency and attaining a level of profitability rather than expanding.[36]

Sweden is amply supplied with iron ore. However, it must import virtually all of its coking coal, much of which comes from the United States.

In terms of trade, Sweden has exported a significant amount of its output during the past decade. In 1969, 1.5 million tons were exported. This increased to 2 million tons by 1974, fell back somewhat in the succeeding years, recovered to 2.2 million tons in 1978, and increased slightly to 2.4 million tons in 1979. Imports, over the same period, have been quite similar in terms of growth and total tonnage, reaching a high point of 2.3 million tons in 1975.[37] Eighty percent of Swedish exports are in the form of specialty steels. Imports are principally carbon-steel products such as cold-rolled sheets.

In regard to technology, Sweden in 1980 had continuous-casting capacity to take care of 49 percent of its raw-steel output. This was a sharp increase from approximately 25 percent in 1975 and will continue to expand further with the new units now being added. In 1980, the oxygen process accounted for 50.7 percent of steel output, with the electric furnace running a close second at 44.4 percent. Some open hearths were still in operation, accounting for about 5 percent of the output, or some 200,000 tons.

During the remainder of the 1980s, Sweden, according to present plans, will concentrate on improving its operations and profitability. Exports will remain at about the same level they attained in 1979 and 1980, and there is no thought of increasing capacity.

Yugoslavia

Steel production in Yugoslavia has doubled in the past fifteen years as output rose from 1.8 million tons in 1965 to 3.6 million tons in 1980. The industry consists of two large integrated steelmaking operations and a number of smaller plants. The two large plants, located at Zenica and Smederevo, have a combined reliable capacity of approximately 4 million tons. The plant at Zenica has just installed a blast furnace capable of producing 2,500 tons a day, while the other plant has a similar furnace nearing completion. There are a number of small electric-furnace operations throughout the country with a total reliable capacity in excess of 1 million tons. Thus, the effective potential to produce steel in Yugoslavia, as of 1982, is approximately 5 million tons. By 1985, it could be as high as 6 million tons. However, Yugoslavia, thus far, has only been able to produce 4 million tons, which was achieved in 1981.[38]

Yugoslavia imports over 2 million tons of semi-finished and 1 million tons of finished steel, at least half of which is in the form of plates for shipbuilding. Exports are small, consisting principally of cold-rolled sheets.

In regard to raw materials, a significant portion of ore and most of the coking coal are imported. In 1980, some 1.8 million tons of ore were imported, principally from South America, while 4.1 million tons of relatively low-grade ore were produced domestically.

The Yugoslavian steel industry will have a minor increase in capacity during the next five years, and thus, it will still have to import steel to care for the needs of the country's economy.

Turkey

The steel industry in Turkey has grown fourfold in the last fifteen years from a production of 600,000 tons in 1965 to 2.4 million in 1980. This growth was achieved through the construction of two new integrated steel complexes, Eregli and Isdemir. These were added to Karabuk, which was already in existence.

In 1980, the capacity of the Turkish industry, which also included a number of small electric-furnace plants, was listed at 4.1 million tons. The rate of utilization was approximately 60 percent for the industry and about the same at the Eregli plant, which has a rated capacity of 1.5 million tons and a production of raw steel, in 1980, of 876,000 tons, or 58 percent of capacity. There were a number of reasons for this low rate of capacity utilization, including unfavorable economic conditions. However, it is interesting to note that, under better economic conditions, crude-steel production at Eregli in 1979 was almost exactly the same as 1980—that is, 875,000 tons, against a capacity of 1.5 million tons. One of the reasons given for the low rate of operation was the difficulty in obtaining and transporting raw materials from their sources to the point of consumption.[39]

Plans for expansion for Turkey are most ambitious as indicated in the 1980 annual report of Eregli. It is expected that, by 1985, 8 million tons of integrated steelmaking capacity will exist in the country. A more-reasonable estimate places the potential for steel production by 1985 between 5 million and 6 million tons, which is an advance over the current 4.1-million-ton rating.

Turkey in the past has been heavily dependent on steel imports. In 1977, 2.2 million tons came into the country. This amount has diminished sharply since that time to 600,000 tons in 1980. Substantial quantities of iron ore and coal are also imported to meet raw-material requirements.

In terms of technology, the basic-oxygen steelmaking process accounts for all of the production of Eregli and Isdemir, which is approximately 60 percent of the country's output. These two plants also employ continuous-casting units.

The future plans, as stated, are indeed ambitious. Turkey's industry will grow in the years ahead but at a slower pace than some had hoped.

Notes

1. Annual Reports of BSC for various years.

2. Interviews with BSC executives.

3. Interview with director general of the British Independent Steel Producers Association.

4. Capacity figures are taken from conversations with corporate officers of the various companies involved, not only in Germany but throughout the EEC.

5. Material on ARBED capacity and construction plans are taken from a publication of ARBED entitled, *ARBED, A Portrait of the Group* (1979).

6. Information on Italsider capacity and facilities was obtained from an interview with the executive personnel of Italsider.

7. IISI, *Annual Statistical Reports, 1981* (Brussels: IISI, 1981), p. 2.

8. Iron-ore production statistics were obtained from the American Iron and Steel Institute, *Annual Statistical Reports,* U.S. Department of the Interior, *Mineral Yearbooks,* and IISI, *Annual Statistical Reports.*

9. Rates are derived from H.P. Drewry, *Statistics and Economics,* (London: December 1981).

10. Ruhrkohle, *Figures Data Facts* (Essen, Germany: Ruhrkohle, 1980).

11. Figures on coal production and international shipments for the EEC countries are taken from *Coal-International* (Washington, D.C.: Zinder-Neris, Inc., February 1980); and *Coal Statistics International* (Washington, D.C.: McGraw-Hill and Zinder-Neris, Inc., February 1981). This latter publication, formerly issued by Zinder-Neris, Inc., is now issued by McGraw-Hill.

12. British Steel Corporation, *Annual Report and Accounts 1979–80* (London: British Steel Corporation), p. 2.

13. IISI, *Steel Statistics Yearbook 1981* (Brussels, 1981), p. 5.

14. "Iron and Steel Scrap, Monthly, December 1979," *Mineral Industry Surveys* (Washington, D.C.: U.S. Department of the Interior, Bureau of Mines).

15. *ARBED, A Portrait of the Group,* a publication of ARBED, S.A.

16. Conversations with executive personnel of Hoogovens.

17. Conversations with executive personnel of DDS.

18. IISI *World Steel Figures 1981,* p. 15.

19. American Iron and Steel Institute, *Annual Statistical Report for 1979,* p. 49.

20. *Monthly Report of the Iron and Steel Statistics, July 1981,* (Japan Iron and Steel Federation, 1981), p. 16.

21. *Metal Bulletin* (London: Metal Bulletin PLC, June 26, 1981), p. 27.

22. William T. Hogan, "World Steel in the 1980's," *Center Lines* (August 1981), p. 3.

23. IISI, *Steel Statistical Yearbook 1981* (Brussels, 1981), p. 2.

24. American Iron and Steel Institute, *Annual Statistical Report for 1938* (New York: 1939), p. 98.

25. IISI, *Steel Statistical Yearbook 1981,* p. 2.

26. Conversations with officials of Voest-Alpine.

27. IISI, *Steel Statistical Yearbook 1981,* pp. 23 and 24.

28. Conversation with executive personnel of Altos Hornos de Vizcaya.

29. IISI, *Steel Statistical Yearbook 1981,* pp. 23 and 24.

30. Internal memorandum on the organization of SSAB.

31. Ibid.

32. SSAB, *Annual Report* (Stockholm, 1978), p. 5.

33. Ibid.

34. Ibid.

35. IISI, *World Steel Figures 1981,* p. 6.

36. IISI, *Steel Statistical Yearbook 1981,* p. 2.

37. Ibid., pp. 23 and 24.

38. Conversations with executive personnel of the Yugoslavian steel industry.

39. Eregli Iron and Steel Works, *Eregli 1980 Annual Report* (Ankara, 1981).

4 Japan Seeks New Markets

The steel industry in Japan was not affected as drastically as that of Western Europe by the steel depression of the late 1970s. Nevertheless, the worldwide decline in demand and production has had a definite impact on Japanese steelmakers.

During the steel boom of 1973–1974, the Japanese integrated companies, in keeping with other producers around the world, inaugurated an expansion program to add a total of 22 million tons to their capacities. Unlike the rest of the world, the Japanese pursued their plans to expand in spite of the sharp drop in demand in 1975 and subsequent years. The plans, which included the construction of seven new blast furnaces, were carried forward and completed in 1978. Five of these furnaces constituted additional capacity, while two were considered replacements. Not until late in 1977, when the Japanese steel industry had not recovered to its boom level and when prospects for the future were not particularly bright, did the industry decide to halt further expansion and concentrate capital investment on improving facilities, conserving energy, and eliminating older equipment.

Policymakers decided that additional capacity would have to await a worldwide growth in demand that would absorb 1978 capacity. One of the industry leaders expressed this philosophy and added, "When demand is equal to supply, we will build a new mill, and this can be done in two years." In keeping with this viewpoint, as of 1982, the Japanese steel industry has no plans to expand; in fact, a program is underway aimed at reducing electric-furnace capacity during the years ahead. It consists of eliminating almost 3 million tons of raw-steel capacity by early 1983 and freezing the installation of new equipment until mid-1983.

Capital expenditures by the five major integrated Japanese producers in 1981 exceeded $3.3 billion. This total amount was designated to rebuild and revamp some existing facilities, as well as to construct additional continuous-casting units, new seamless-tube mills, new coating lines, and new annealing lines.

Programs to increase seamless-tube facilities include three new mills, one each at Nippon Steel, Nippon Kokan, and Sumitomo, while Kawasaki will upgrade two of its existing facilities. Total seamless-pipe capacity to be added from these installations will be 1.8 million to 2 million tons, directed at the expanding world market for oil-country goods.

New continuous-casting facilities will be constructed, while others will be upgraded to handle larger tonnages. In 1980, the Japanese industry cast about 60 percent of its raw-steel production, an increase from 46 percent in 1978. In terms of tonnage, continuous casting rose from 47.2 million in 1978 to 66.3 million in 1980.[1] This was the result of the start-up in 1979 and 1980 of twenty new continuous-casting machines, five of which were designed for slabs and the remaining fifteen for billets.

From 1981 to 1983, an additional thirteen continuous-casting units will be installed, bringing the amount of continuously cast steel by the Japanese industry to at least 70 percent. Nippon Steel will have 75 percent of its output continuously cast, while Kawasaki will reach 90 percent.

Growth

The Japanese steel industry in the postwar period experienced the most remarkable growth of any steel industry in history. From an output of less than 1 million tons in 1947, it rose to 119 million tons in 1973. The major portion of this was achieved in the 1960s, as production rose from 22.1 million tons in 1960 to 93.3 million tons in 1970.[2] In the 1970s, additional capacity was installed so that, by 1980, the industry had a production capability of 140 million to 145 million tons. The growth had a twofold purpose:

1. To serve the expanding needs of the Japanese domestic economy that was growing rapidly in the 1960s and early 1970s,
2. To serve the world market for steel that was expanding rapidly during the same period.

To finance this huge investment, the industry depended heavily on borrowing. This was made possible, according to a leading Japanese banker, because the lenders recognized that "the steel industry was basic to the nation's economy and to the strong confidence they placed in the growth potential and safety of the industry as an investment." Thus, "the nation's private financial institutions continuously met the huge investment-fund requirements of the steel industry whose equity ratio was at a low level and whose earning power could hardly be rated sufficient."[3]

The rapid increase in capacity was aided considerably by the relatively low cost per ton of raw-steel capacity installed. In the latter half of the 1960s, it was approximately $110 per ton. This cost level was maintained, in part, by the fact that a small amount of antipollution investment was involved.

When asked what type of security the lenders required, a Japanese banker replied:

The reason the Japanese steel companies were able to obtain necessarily large loans was because the Japanese steel industry had the tremendous zeal in expanding its production as well as the extraordinary confidence about improving its productivity and strengthening its international competitiveness. . . . It was this strong aspiration and self-confidence which prompted the industry to daringly and successfully bring those hesitating and foot-dragging Japanese bankers to their side. And the result was that they achieved a great success in doing so.[4]

The last expansion program, inaugurated in 1973 and 1974, was based on the conviction that world demand would remain strong and thus would require Japanese exports. The certitude that this demand would continue was stated in 1973 at a worldwide meeting of the steel industry:

Judging from the present world demand and supply situation of iron and steel, while demand is expected to grow steadily both in the advanced and developing countries throughout the 1970s, an increase in the supply capacity of advanced steelmaking countries and an improvement in the capabilities of developing countries to meet their own demand will still take some time to materialize, and so the world demand for Japanese steel exports is expected to remain strong.[5]

In view of these projections, capital was readily available for the expansion that took place between 1974 and 1978, resulting in the construction of a substantial amount of capacity that proved to be in excess of demand due to the worldwide steel depression that began in 1975. The expansion program, as indicated, involved seven new, large blast furnaces, each capable of producing 9,000–10,000 tons of iron per day. Five of these were placed at existing plants, where they joined similar-sized units; two were built at the new Nippon Kokan Ohgishima plant, constructed on filled-in land near Tokyo. This plant was principally a replacement of other facilities and partly an expansion.

Large steelmaking plants with huge blast furnaces are characteristic of the Japanese industry. No country in the world has as many large steel plants as Japan. Further, of the sixteen blast-furnace plants operated by the six integrated companies, ten have an annual capacity to produce 8 million tons or more of raw steel. Among these, the Nippon Kokan plant at Fukuyama City is the world's largest, with an annual capacity of 16 million tons of raw steel. All of the blast-furnace plants in Japan are located on the seacoast with deep-water ports. This, if not necessary, is highly advantageous, since the industry must import virtually all of its iron ore and about 90 percent of its coal. These raw materials amounted to more than 180 million tons in 1979 and reached a high of 192 million in 1980. They were brought to Japan in large bulk-cargo carriers with capacities, in the case of ore boats, of up to 150,000 tons. Coal often came in smaller ships because of the relatively shallow ports in the exporting countries where it was loaded. Once the imported raw materials are unloaded, they are at their destination and no additional transportation is required.

The seacoast location also affords an advantage in exporting steel. The Japanese have exported from 30 million to 37 million tons of steel annually during the past ten years, which tonnage has moved directly from seacoast plants to the country of destination.

The electric-furnace segment of the Japanese steel industry expanded from an output of 1.2 million tons in 1955 to 15.6 million tons in 1970 and then to a record 27.3 million tons in 1980. Between 1978 and 1980, electric-furnace steel output increased by 5 million tons. The bulk of this increase, 4 million tons, came in 1979, as overall steel production increased by more than 9 million tons.[6]

The increase in electric-furnace output was not so much the result of additional furnaces but, rather, the increase in the size of some furnaces and the improvement in operating techniques. The capacity of electrical transformers was increased, oxygen burners were introduced, water-cooled side panels were employed, and improvements were made in material handling. In the next two years, as previously indicated, a number of electric furnaces will be scrapped, reducing the capacity of this segment of the Japanese steel industry by 3 million tons.

During the next five years, the Japanese industry's raw-steel capacity will remain at about the same level that it was in 1981—that is, 140 million to 145 million tons. No new blast furnaces will be built to add to capacity, although one or two will probably be replaced like the one recently announced by Nippon Steel for its Muroran Works. Originally there were four small furnaces, by Japanese standards, at this plant. Numbers two and three have been taken out of service, and number one, which is a 1,250-cubic-meter furnace, will be replaced by a new 2,500-cubic-meter furnace that will not only replace number one but will make up, in part, for the capacity lost when numbers two and three were shut down. The new furnace is scheduled to be blown in during December of 1983 or January 1984.[7]

During 1980 and 1981, one-third of the Japanese blast furnaces were idle and banked. Thus, with this much capacity in reserve, there is no incentive whatever for the integrated producers to add new facilities. Nevertheless, the industry will increase its ability to ship steel products through the improvement of existing facilities, as well as the further installation of continuous casting, which provides a greater yield from raw steel to the finished product. This will permit shipments to rise, although basic raw-steel capacity will remain stable.

In 1979 and 1980, Japanese steel production was 111.7 million and 111.4 million tons respectively. This could be increased to 135 million tons in a relatively short period should demand require it. It would not be possible, however, for the Japanese to reach the theoretical capacity of 140 million to 145 million tons without investing substantial sums in other facilities throughout their plants where bottlenecks exist. One company, with a rated theoretical capacity of 20 million tons and a recent annual production in the area of 12 million to 13 million tons, reports that it could not produce up to its rated capacity without substantial capital investments. The management feels certain, however, that if there were an increase in demand it could reach 16 million tons without any difficulty.[8]

The finishing capacity of the Japanese steel industry is adequate to absorb 135 million tons of raw-steel output. The Japanese certainly could add 25 million tons to the 1979 and 1980 production figures, if it were warranted by demand.

At the present level of world steel demand, Japan obviously has excess capacity, a condition that has existed since 1975. In 1977, this prompted a high official in the industry to state that capital was too readily available in the 1960s and 1970s so that the Japanese steel industry, taking advantage of this situation, overbuilt. This, in a sense, is hindsight, since the projections of the early 1970s indicated a continued growth in world steel demand. Unfortunately, this demand did not materialize during the latter part of the 1970s.

Should world steel demand increase annually by even a modest 1.5 percent during the next five years, the Japanese would be in a position to take full advantage since they have idle capacity that can be put into operation on short notice. In slack times, however, there will be resistance in a number of countries throughout the world to Japanese exports. The Japanese are fully aware of this and, thus, will not expand their capacity beyond its current level during the first half of the present decade. In the second half of the decade, it is improbable that they will add much capacity because of the expansion plans of Third World countries that currently import significant quantities of Japanese steel.

Raw Materials

Japan is a resource-poor nation and must import almost all of the raw materials required for steelmaking. In 1980, it imported 133.7 million tons of ore, which was over 99 percent of its requirements. In the same year, 58.3 million tons of coking coal, or 90 percent of its requirements, were imported.[9]

The raw-material needs have increased proportionately as the industry grew during the past two decades. In 1960, a production of 22.1 million tons of steel, 15 million tons of which came from the open hearth and only 2.6 million from basic-oxygen converters, required imports of 13.4 million tons of ore. By 1965, the industry had doubled its output and made a dramatic shift to the basic-oxygen process, which accounted for 22.6 million tons as opposed to 10.2 million for the open hearth. In that year, some 35.2 million tons of ore were imported to provide feed for blast furnaces, which supplied the molten pig iron for the oxygen process.[10]

By 1970, steel production had more than doubled over the 1965 figure, with the basic-oxygen converter accounting for 73.8 million tons out of a total of 93.3 million. Blast-furnace output to feed the oxygen converters amounted to 67.5 million tons, requiring 83 million tons of iron-ore imports.[11] The operation of the basic-oxygen process in Japan employs a practice that uses more molten pig iron than the industries in other countries. In the United States, the charge into the basic-oxygen converter includes 25–30 percent scrap. This is also true in a number of European countries. However, in Japan, the precentage of scrap in the charge is closer to 15 percent and, at times, has dropped to as low as 7 percent. Thus, more iron is necessary, and consequently, more ore is needed to produce the iron.

The Japanese decided, during the growth period in the 1960s, to invest heavily in large blast furnaces. This was done in spite of the fact that the country has virtually no iron ore and was completely dependent on outside sources for this essential iron-making material. They had no alternative because plans to expand the industry involved the installation of the new oxygen steelmaking process which must be supported by blast furnaces. It would not have been possible to produce the tonnages envisioned by the electric-furnace method, and the open hearth was rapidly becoming obsolete.

Ore

Fortunately for the Japanese, there were a number of discoveries of large, rich deposits of iron ore in various locations throughout the world between 1950 and 1970. These included extensive deposits in Australia, Venezuela, Brazil, Peru, Canada, India, West Africa, and South Africa. Thus, although the dependence of the Japanese on outside ore was complete, the sources were multiple and the dependence did not have to be concentrated in terms of geographic location. In the early 1960s, as can be seen from the following list, the supply of iron ore was obtained from a number of areas:

Southeast Asia	38.5%
South America	27.0%
India	20.3%
North America	11.0%
Africa	3.1%[12]

Southeast Asia included the Malay states and the Philippines. South America included Peru, Chile, and Brazil, with the major portion coming from Peru.

In the 1960s, extensive reserves of high-grade ore were found in Australia. By 1965, a number of these areas were being developed with capital made available based on long-term ore contracts with the Japanese; in fact, it could be said that the Australian iron-ore industry got its start from the Japanese commitments to take large tonnages over protracted periods. The first deliveries from Australia came in 1965, after which they grew rapidly, reaching 28 percent of Japanese imports in 1969 and 36 percent in 1970. By 1973, the year of Japan's record output, Australia provided 47.6 percent of its iron-ore requirements. Throughout the 1970s, Australia maintained its position, while India, though increasing its tonnage, declined in its relative position. South America, particularly Brazil, has come on strong and provides a substantial tonnage. Table 4–1 gives the major sources of iron-ore imports for selected years from 1962 to 1979. It should be noted that, although a number of countries have been significant suppliers from time to time during the twenty-year period, they have not sustained a large tonnage and so were not included in the table; among these were Canada and the United States.

In the late 1960s, the United States shipped an average of 3 million tons a year. However, in 1972, this figure was cut drastically as the pellet contract with Kaiser Steel ran out. After 1974, the United States did not ship ore to Japan. Canada supplied an average of less than 2 million tons a year in the 1960s, increasing this to between 3 million and 4 million tons a year in the 1970s.

Since Japan is devoid of iron ore, it must continue to import its full requirement. The question is from where, and the answer seems to be evident—Australia and Brazil will dominate, India will be in third place, and other countries will follow at a distance.

Coal

The Japanese dependence on imported coking coal has grown to about 90 percent of its requirements. In 1960, with the production of 13.4 million

Table 4-1
Major Sources of Japanese Ore Imports for Selected Years: 1962-1979
(millions of metric tons)

Year	Total Tonnage	Percentage of Total			
		Australia	*India*	*Chile and Peru*	*Brazil*
1962	22.1	—	20.3	24.0	2.2
1965	38.8	0.5	19.9	28.0	2.0
1967	56.4	14.8	18.8	26.4	4.3
1969	83.1	28.0	16.3	19.7	5.0
1970	102.0	35.9	16.1	15.5	6.7
1971	114.8	40.3	14.6	14.3	7.8
1972	111.4	43.3	16.0	12.2	8.3
1973	138.7	47.6	14.2	10.5	9.5
1974	141.8	47.9	12.2	10.3	13.8
1975	131.7	48.1	12.8	8.0	17.8
1976	133.7	47.9	13.2	NA	19.0
1977	132.6	47.6	13.5	NA	17.9
1978	114.9	45.9	12.5	NA	18.2
1979	130.3	42.4	13.1	7.4	20.1

Source: Japan Iron and Steel Federation, *The Steel Industry of Japan* for various years.
NA means not available.

tons of pig iron, coke requirements were 8.1 million tons. To produce this, Japan imported 5.7 million tons of coking coal to supplement 5.5 million tons from domestic sources. This was the first year that imports exceeded domestic supply. From that time forward, metallurgical-coal imports grew rapidly, while domestic production reached a peak of 10.4 million tons in 1971 and declined to 6.9 million tons in 1980. Table 4-2 lists the tonnages for imported and domestically produced coking coal from 1960 to 1980.

The foreign sources of coking coal are principally three—the United States, Australia, and Canada. In the early 1960s, the main source by far was the United States, accounting for almost 60 percent of Japanese imports, and Australia provided the second largest amount with 27 percent in 1962. For the remainder of the decade, the United States maintained its dominant position. However, this dominance began to change as Australia increased its coking-coal exports and, in the early 1970s, replaced the United States as the number-one supplier.

The development of coal resources in Canada brought this nation into the picture in the early 1970s, and since then, it has achieved a significant participation. Table 4-3 lists the shares of Japanese coking-coal imports

Table 4-2
Imports and Domestic Production of Japanese Coking Coal: 1960-1980
(millions of metric tons)

Year	Imported	Domestic	Total	Imports as Percentage of Total
1960	5.7	5.5	11.2	50.9
1961	8.2	6.5	14.7	55.8
1962	9.3	6.4	15.7	59.2
1963	9.3	7.5	16.8	55.4
1964	11.2	8.5	19.7	56.9
1965	14.3	8.2	22.5	63.6
1966	16.5	9.3	25.8	64.0
1967	22.0	10.1	32.1	68.5
1968	27.7	9.3	37.0	74.9
1969	37.1	9.4	46.5	79.8
1970	44.4	9.4	53.8	82.5
1971	44.9	10.4	55.3	81.2
1972	43.8	10.0	53.8	81.4
1973	53.7	10.2	63.9	84.0
1974	56.8	9.2	66.0	86.1
1975	57.7	7.6	65.3	88.4
1976	56.0	7.2	63.2	88.6
1977	56.3	6.5	62.8	89.6
1978	49.0	6.7	55.7	88.0
1979	51.8	6.5	58.3	88.9
1980	58.0	6.9	64.9	89.4

Source: Japan Iron and Steel Federation, *Statistical Yearbook,* 1972, p. 15; Japan Iron and Steel Federation, *Monthly Report of the Iron and Steel Statistics* (July 1981), p. 19.

accounted for by the three major sources during selected years from 1962 to 1980.

In terms of providing coking coal to Japan, the USSR accounted for 10.1 percent of Japanese imports in 1962. This figure declined to 7 percent in 1965, and in subsequent years, combined statistics for the USSR and Poland indicate a participation of between 6 and 7 percent.

The Chinese have made a bid to become a factor as a source of Japanese coking coal. In 1978, Japan imported 300,000 tons, which will be increased annually up to 1982, at which time it will reach 2 million tons. At that time, a decision will be made as to whether or not to increase the level of imports from China above 2 million tons.

Table 4-3
Japanese Imports of Coking Coal for Selected Years: 1962-1980
(percent of total)

Year	Australia	United States	Canada
1962	27.0	56.4	4.8
1965	41.3	43.8	5.1
1967	36.7	43.1	3.6
1969	38.2	48.8	1.2
1970	32.0	53.4	6.9
1971	35.7	41.4	14.8
1972	40.7	35.5	16.3
1973	43.3	30.5	19.1
1974	35.1	42.3	15.4
1975	38.8	38.5	18.1
1976	42.7	30.5	17.6
1977	44.0	26.9	18.5
1978	47.9	18.2	22.1
1979	45.7	24.8	18.9
1980	41.4	31.9	17.2

Source: Japan Iron and Steel Federation, *Monthly Report of the Iron and Steel Statistics,* for the years involved.

In the years from 1975 through 1980, Japan imported approximately 89 percent of its coking-coal requirements. A change has occurred, however, in the quality of the coal imported. In the middle 1970s, the Japanese used approximately 18 percent of low-volatile coal in the charge for the coke ovens, most of which was obtained from the United States. Since 1978, there has been a sharp decrease in the amount of low-volatile coal to the point at which it now constitutes about 7 percent of the coking-coal mix that is charged into the ovens. Further, the Japanese have developed a process that involves briquetting a significant part of the charge. The coal briquettes, which are used as a partial charge, are formed from a combination of low-grade coking coals and noncoking coals. This process is used extensively in Japan and reduces the country's dependence on more-expensive, higher-grade coking coals.

There is no question that Japan will continue to import a very high percentage of its coking-coal requirements for the remainder of the decade. This is particularly true since the other technological development that is on the horizon—namely, formed coke using inferior-grade coals—will not be in commercial operation for at least a decade.

Scrap

The Japanese steel industry is particularly notable for its use of the basic-oxygen process. However, as output grew between 1960 and 1980, the electric furnace assumed a highly significant role in steel production, thereby increasing the demand for scrap. In 1953, electric-furnace output passed 1 million tons for the first time. It increased steadily over the next few years so that by 1960 it accounted for 4.5 million tons of production. In 1965, it reached 8.4 million tons; in 1970, 15.6 million tons. In 1973, the year of peak Japanese production, it accounted for 21.4 million tons. During the next few years, with the overall decline, electric-furnace production decreased slightly but surpassed its former record in 1978 with 22.4 million tons. During the next two years, it experienced a remarkable increase as output rose by over 4 million tons in 1979 and approximately an additional 1 million in 1980.[13]

The electric-furnace segment of the Japanese steel industry has been heavily dependent on scrap imports, although this dependence has been declining somewhat in recent years. In 1970, scrap imports amounted to 5.8 million tons, of which 4.6 million tons came from the United States. In 1973, with a significant increase in electric-furnace production of almost 6 million tons, total imports were slightly less than in 1970, amounting to 5.4 million tons, of which 4.4 million came from the United States. In the next few years, imports of scrap dropped significantly to a low of 1.4 million tons in 1977 as electric-furnace production declined to 19.6 million tons. With the resurgence of output from electric furnaces, scrap imports again increased and, in 1980, amounted to 3 million tons, of which 2.6 million tons came from the United States.[14]

These statistics are quite significant when one considers that in 1974, with an output of 20.9 million tons of electric-furnace steel, scrap imports were 5.4 million tons. In 1980, with production rising to 27.2 million tons, imports stood at 3 million tons. Clearly, Japan is generating more scrap and has become less dependent on external sources. Much of this home scrap comes from the iron and steelworks that produce significant quantities as raw steel moves through the various stages of production to the finished product. Further, as indicated, the Japanese use a very high percentage of pig iron in their basic-oxygen-furnace charge and relatively little scrap. In 1980, for example, the oxygen converters consumed 7.7 million tons of scrap to produce 84.1 million tons of steel. This compares with 10.4 million tons to produce 84.4 million tons of oxygen steel in 1975. Table 4–4, which gives total tonnage of scrap imports, as well as the amount from the United States and the output of electric-furnace steel, indicates the decreasing dependence of Japan on imported scrap during the past twenty years.

The Japanese economy will generate more scrap in the next decade

Table 4-4
Japanese Imports of Scrap for Selected Years: 1960-1980
(millions of metric tons)

Year	Total Import Tonnage	Imports from the United States	Electric-Furnace Production
1960	4.4	3.1	4.5
1961	7.0	5.3	5.9
1963	4.6	3.4	7.3
1964	5.1	3.8	8.4
1966	3.5	2.7	9.2
1967	6.7	5.0	11.4
1969	4.9	3.7	18.7
1970	5.8	4.6	15.6
1971	2.5	1.8	15.6
1972	2.5	1.9	18.0
1973	5.4	4.4	22.4
1974	3.6	2.5	20.9
1975	3.1	2.4	16.8
1976	1.8	1.1	20.0
1977	1.4	0.9	19.6
1978	3.2	2.7	22.4
1979	3.3	2.7	26.4
1980	3.0	2.6	27.3

Source: Japan Iron and Steel Federation, *Monthly Report of the Iron and Steel Statistics,* for the years involved.

since it has been using large tonnages of steel for the last fifteen to twenty years. Much of this steel has gone into products and facilities that are rapidly becoming obsolete and will be scrapped. Further, the manufacturing segment of the economy that produces products made of steel generates a significant amount of scrap in the manufacturing process, which is called prompt-industrial scrap. Clearly, an automotive industry that produced 11 million vehicles in 1980 will generate much more prompt industrial scrap than was produced as recently as 1966 from a production of 2.3 million vehicles. In spite of these increases, Japan will continue to import scrap for the remainder of the decade, but the tonnage will decline.

Transportation of Raw Materials

The quantities of raw materials, which in 1979 totaled over 185 million tons and in 1980 over 195 million tons, had to be transported to Japan over long distances from Brazil, Chile, Peru, the east coast of the United States, Canada, Australia, India, and South Africa. Estimates indicate that, in 1979, total bunker-oil consumption needed by ships to move these materials to Japan was about 5 million tons. This amounted to approximately 60 kilograms of bunker oil per ton of pig iron produced in that year.

Before the oil crisis of 1973, bunker oil cost approximately $15 per ton. In 1973, although prices differ somewhat in various parts of the world, it rose to $80 per ton; in 1980, to $180; and by 1981, $220 a ton.[15] In 1981, this amounted to a total cost of approximately $1.1 billion and added between $10 and $11 per ton to the cost of Japanese steel production.

In order to relieve this situation and to hold down increases in freight costs, an effort is being made to develop new types of energy-saving ships with an emphasis on economy rather than speed. A carrier is under construction in Japan, to be launched in 1982, that will reduce bunker-oil consumption from 14 kilograms per ton, the amount now used to haul ore from Australia to Japan, to 10 to 11 kilograms. Another ship, to be launched after 1982, is designed to reduce the present 14 kilograms to 7 kilograms. These are only two ships, and for the program of cost reduction to be effective, many more will be needed and will take time to build.

Another effort to conserve fuel that Japan must import was to take oil off the blast furnace as a partial fuel and replace it with coke and, in some cases, with tuyere-injected, pulverized coal. In the past, some of the Japanese blast furnaces used oil to provide from 10 to 20 percent of the fuel needed to reduce iron ore to pig iron. With the reversion to a total coke charge, 10 to 20 percent more coke will be necessary. By mid-1982, it is anticipated that nearly all of Japan's large blast furnaces will be operating completely on coke.

The increase in the price of transportation due to rising oil prices has meant additional costs for Japanese raw materials. In the 1960s and early 1970s, the availability of cheap raw materials, due to the large bulk-cargo carriers in which they were transported, gave Japan a definite cost advantage. This advantage is diminishing with the increase in oil prices, and it will be some time before enough ships can be built or converted to use less oil and to reduce transportation costs. The Japanese are constantly seeking more ways to reduce energy costs by improving efficiency. This search becomes more urgent as oil prices increase. The Japanese once had a significant cost advantage in terms of assembling raw materials; they now see this slipping away.

Many have considered Japan's steel industry vulnerable because of its dependence on imported raw materials, and the increased concentration of its imports from Australia that has developed in the last two decades would seem to increase this vulnerability. Such may appear to be the case, but upon further examination, it becomes clear that Australia, the main supplier of Japanese ore and coal, is also heavily dependent on Japan for its market. The sale of these raw materials in 1979 brought revenues to Australia in excess of $3 billion, and it would be a serious blow to its economy if this revenue were sharply reduced. Therefore, the dependence is mutual between Japan and Australia. Japan's steel industry would be crippled without Australian raw materials, and the Australian economy would be seriously injured without the Japanese market.

Technology

The Japanese steel industry experienced the major portion of its growth during the 1960s, as it moved from 22.1 million tons of raw steel in 1960 to 93.3 million tons in 1970. This overshadowed a significant growth in the 1950s from 4.8 million tons in 1950 to 22.1 million tons in 1960. It is interesting to note that during the 1950s the open hearth was the principal contributor to total steel output, since production from this process rose from 3.6 million tons in 1950 to 15 million tons in 1960. In the 1960s, large tonnages of oxygen steel capacity were installed.

Oxygen Steelmaking

The oxygen converter was first installed in the mid-1950s on a very modest scale. By 1957, its output was 454,000 tons, growing to 2.6 million tons by 1960. Fortunately for the Japanese, the decade of its major growth achievement came at a time when the basic-oxygen process had been in operation for a sufficient number of years to prove its worth. By 1960, it was accepted as a well-developed, efficient operation. Consequently, any company adding steel capacity in a fully integrated plant operating coke ovens and blast furnaces turned to oxygen converters. Unfortunately for the U.S. steel industry, its expansion came in the early 1950s before the oxygen process was fully proven.

The oxygen process was responsible for the major portion of steel output in Japan after 1964. In that year, with a production of 39.8 million tons, it accounted for 17.6 million tons, or almost half, while the open hearth dropped to 13.9 million tons and the electric furnace contributed 8.4 million tons.[16] In the following year, 1965, the oxygen process produced 55 percent of Japanese steel. This process increased in tonnage and as a percentage of output, reaching its peak in 1973 with 96.1 million tons, or 81 percent of total production.

The initial installations in the late 1950s and early 1960s were small units, many of them with less than 100-ton capacity per heat. In 1965, with twenty basic-oxygen steelmaking shops in operation, thirteen of these operated converters with less than 100-ton-heat size, and of the remaining seven, five had converters between 150 and 200 tons per heat. In the late 1960s and 1970s, the smaller converters were replaced with larger ones, some of which were capable of producing heats in excess of 300 tons and many of which were in the 250–300-ton range.

The Japanese industry now produces all of its steel by the oxygen-converter or electric-furnace process, the open hearth having been eliminated from the statistics in 1978. There are no current plans to expand oxygen-steelmaking capacity in Japan because it is producing at about 70 to 75 percent of its potential; thus, no incentive exists to add new units.

Continuous Casting

The first continuous-casting unit in Japan was installed in 1960. Others followed in 1964 and 1965, all of which, with one exception, were dedicated to casting alloy or stainless steel. Not until 1967 was a wide-slab caster installed by Nippon Kokan that was capable of producing carbon-steel slabs up to 64 inches in width. From 1967 through 1969, fifteen continuous-casting units, most of which produced billets and blooms, were installed, bringing the industry's total number to twenty-one units. These allowed Japan to cast 3.3 million tons of steel, or 4 percent of its total 1969 output.[17]

In the 1970s, this output rose rapidly both in terms of tonnage and percentage of total. By 1974, raw steel cast amounted to 29.4 million tons, or 25 percent of total production. Table 4–5 indicates the growth from 1969 through 1980, when 66.3 million tons, or 59.5 percent of steel production, was continuously cast.

According to investment plans for the early 1980s, additional continuous-casting units will be installed, bringing the total for the country up to at least 70 percent of its raw-steel production; by 1985, this figure could conceivably be 75 percent. Eleven continuous-casting units are currently on order or under construction; two others were installed in 1981.

The Japanese are by far the leaders in the amount of tonnage continuously cast. In terms of the relative amount cast, other nations with little steel production, such as Denmark and Finland, have a higher ratio of cast steel than the Japanese; however, their tonnage is quite limited. In the case of Denmark, the 500,000 tons cast represent 73 percent of output, and in Finland, the 2.3 million tons cast represent 90 percent.

Table 4–5
Continuously Cast Steel Output in Japan: 1969–1980
(millions of metric tons)

Year	Tonnage Cast	Percentage of Raw-Steel Output
1969	3.3	4.0
1970	5.3	5.6
1971	10.0	11.2
1972	16.5	17.0
1973	24.7	20.7
1974	29.4	25.1
1975	31.8	31.1
1976	37.6	35.0
1977	41.8	40.8
1978	47.2	46.2
1979	58.1	52.0
1980	66.3	59.5

Source: International Iron and Steel Institute, *Statistical Reports* for the above years (Brussels).

Continuous casting has contributed to energy saving as well as a higher yield from raw steel to the finished product and permits the Japanese to ship more steel in relation to raw steel output than any other major steel-producing nation.

Coke Production

Since Japan is dependent on imports for almost 90 percent of its coking coal, it has striven to reduce this dependence by developing new processes for producing coke that, although they still require imported coal, are capable of functioning adequately with less costly lower grades of coal. The principal development in this respect has been a process in which a portion of the coke-oven charge is briquetted, thus increasing its density and permitting the production of quality coke with lower-grade coals. The coal briquettes that are used are formed from a combination of low-grade coking coals and, in some instances, non-coking coals. These briquettes constitute 30 percent of the total charge, with the remaining 70 percent made up of higher grade, finely ground coal.

Briquetted Coal

Two briquetting processes are in operation in Japan, and they differ significantly. One of these, developed by Nippon Steel, uses a coke-oven charge made of a mixture of approximately 20 percent low-grade coking coal and 80 percent good coking coal. Some 30 percent of the entire mixture is briquetted, and the remainder is charged as fine coal. The briquettes have a much higher density than the coal, registering 1,150 kilograms per cubic meter as opposed to 700 kilograms per cubic meter for fine coal. The mixture adds strength to the coke and improves its high-temperature performance in the blast furnace.

The briquettes are formed by the addition of pitch, which acts as a binder. Thus formed, they are blended with fine coal and charged into the oven. The resulting coke product is satisfactory for use in the large Japanese blast furnaces and allows the ratio of higher-grade metallurgical coal to be reduced from 70 percent to 50 percent.

Nippon Steel claims that the foremost result of its process is that the use of U.S. coal, which has the best coking properties among the world's hard coking coals, can be substantially reduced. This is perhaps an overstatement; however, the amount of U.S. coking coal, particularly low-volatile coal used in Japan, has been reduced. The proponents of the process claim that the briquetting will produce additional coke strength over the standard coking process using the same coals and equal strength using inferior coals.

The other process, developed by Sumitomo, is called Sumi-Coal. This consists of a mixed charge of briquettes and fine coal, with briquettes again representing 30 percent of the total. The basic difference is that the briquettes are made of a different mixture than the fine coal. The fine coal, constituting 70 percent of the charge, is good coking coal, although some of it can be a lower grade. The briquettes are made of 70 percent noncoking coal, and the remainder is coking coal. Since the briquettes constitute 30 percent of the mix, the resultant amount of noncoking coal in the entire mixture is approximately 20 percent.

Sumitomo states that its Sumi-Coal system offers a number of advantages including:

An increase of coke strength using the present coal blend;

The use of poor coking coal and/or noncoking coal in place of hard coking coal, at least in part;

The fact that any kind of bituminous binder such as coal-tar pitch or petroleum pitch can be used;

The easy application of the process to existing coke ovens;

Low investment since substantial modifications are not necessary.

At present, Japanese capacity for briquetting coal is about 20,000 tons per day. Facilities are in operation at three Nippon Steel plants, two Nippon Kokan plants, two Sumitomo plants, and one Kobe plant. The future development of briquettes will depend on the availability of coking coal and its price. Leading Japanese authorities on the subject seem to think that briquetted coal has reached a point in its development at which it will level off. Thus, it does not appear that more plants for briquetted coal will be constructed in the near future. It is estimated that the coal used in the briquettes would have to be $12 to $15 per ton less than the other coking coals used in order to justify the application of this process. Some of the coal for briquettes is imported from South Africa, and as might be expected, the tonnage from this source has risen sharply in the last few years to over 3.1 million tons in 1980.

A second major technological breakthrough whereby inferior coals are substituted for high-quality metallurgical coal in the coking process is formed coke. This is a generic name for the resultant material made from both inferior coking coal and noncoking coal and, in some instances, subbituminous coal. The Japanese have had two pilot plants in operation for some time and have produced enough formed coke to make some blast-furnace tests. The results have varied considerably and were not considered successful enough to pursue the process to a production stage. Some consideration was given to this process recently by Sumitomo, but with increase in the price of steam coal, as well as the risk involved, the project has been abandoned.

The Japanese industry has made great strides in adopting and improving on steel-industry technology. This is particularly true in the large plants that have been constructed on its seacoast. A plant capable of producing 10 million tons of steel has a huge materials-handling problem. The ore used in one year is approximately 15 million tons, while coal amounts to approximately 8 million tons. To manage this tonnage, materials-handling facilities have been developed to an extraordinarily proficient degree. Further, the large blast furnaces, capable of producing 10 thousand tons of iron per day, are definitely a Japanese development. They have thirty of these, and their blast-furnace operations have resulted in a very low coke rate that is the envy of the world steel industry. In the steelmaking and rolling sectors of the industry, numerous innovations have been developed to improve the quality of the steel, to increase productivity, and to reduce costs. These range from heat conservation at the basic-oxygen converter to continuous-annealing lines for production of steel sheets and the application of computer controls in most of the steelmaking and finishing operations.

Since 1978, when the industry stopped expanding, it has concentrated its efforts on installing technology to improve efficiency, reduce costs, and save energy.

Trade

In the prewar as well as the early postwar years, Japan was not a steel-exporting nation of any significance. This was due in part to the size of its production that, during those two periods, was quite limited. In the prewar period, crude-steel production in Japan ranged from 3.8 million tons in 1934 to a high of 6.9 million tons in 1940. Very little of this was available for exports, which only twice, in 1936 and 1939, surpassed the million-ton mark in terms of raw-steel equivalent. In the immediate postwar period, Japanese steel production was drastically cut, so that it was 1950 before it reached approximately 5 million tons, and with the pressing needs of the country during those years, little was available for export.[18]

In the years that followed, particularly between 1965 and 1980, Japan became the world's leading exporter of steel so that the very existence of its industry as it is structured depends on a large volume of exports. In the 1950s, exports were conceived as a means of paying for the raw materials that had to be imported for the industry to function. In 1965, when Japan's steel production of 41 million tons made it the third largest producer in the world, it had to import 35.2 million tons of iron ore and 14.3 million tons of coking coal. At this point, it was evident that if the Japanese industry was to continue to expand, very large amounts of raw materials had to be brought in and paid for by developing a more-extensive export market for steel.

The industry made a decision in the 1960s to move heavily into the growing world market for steel and thus justified the construction of huge seacoast plants, where imported raw materials could be turned into steel for export. The Japanese domestic demand for steel could never have justified the size of the industry that was built in the 1960s and early 1970s. It was designed to produce steel not only for Japan but for the world.

This growth in exports by 1976, which rose from 2.5 million tons of products, or 3.1 million tons of raw-steel equivalent, in 1960 to 37 million tons of products, or 42.4 million tons of raw-steel equivalent, was an outstanding achievement made possible by the adoption of the new trade philosophy. Table 4–6 gives the exports of steel products from 1960 to 1980 and indicates the point of departure, in the mid-1960s, when exports became a way of life. Table 4–7 gives the raw-steel production and exports in terms of raw-steel equivalent for the same years.

In 1960, the Japanese exported 14 percent of their output; by 1976, this had increased to 39.5 percent in terms of raw-steel equivalent. This figure is substantial, but it does not represent the entire picture for, in addition to the steel industry, a number of other large Japanese industries, such as shipbuilding and automotive, are very heavily export oriented. In the late 1960s and early 1970s, when the shipbuilding boom was at its height, Japan produced half of the world's tonnage in the form of medium- and large-size

Table 4-6
Japanese Exports of Steel Products: 1960-1980
(millions of metric tons)

Year	Tonnage	Year	Tonnage
1960	2.5	1971	24.2
1961	2.5	1972	22.0
1962	4.1	1973	25.6
1963	5.6	1974	33.1
1964	6.9	1975	30.0
1965	9.9	1976	37.0
1966	9.1	1977	35.0
1967	9.9	1978	31.6
1968	13.2	1979	31.5
1969	16.6	1980	30.3
1970	18.0		

Source: Japan Iron and Steel Federation, *Monthly Report of the Iron and Steel Statistics,* for the years involved.

Table 4-7
Japanese Raw-Steel Production and Exports: 1960-1980
(millions of metric tons)

Year	Raw-Steel Production	Exports[a]	Percentage of Production Exported
1960	22.1	3.1	14.0
1961	28.3	3.2	11.3
1962	27.5	5.3	19.3
1963	31.5	7.2	22.9
1964	39.8	8.9	22.4
1965	41.2	12.7	30.8
1966	47.8	12.2	25.5
1967	62.2	11.3	18.2
1968	66.9	16.3	24.4
1969	82.2	19.9	24.2
1970	93.3	22.3	23.9
1971	88.6	28.3	31.9
1972	96.9	26.0	26.8
1973	119.3	30.2	25.3
1974	117.1	38.4	32.8
1975	102.3	34.4	33.6

Table 4-7 continued

Year	Raw-steel Production	Exports[a]	Percentage of Production Exported
1976	107.4	42.4	39.5
1977	102.4	39.5	38.6
1978	102.1	35.9	35.2
1979	111.7	35.2	31.5
1980	111.4	34.1	30.6

Source: Japan Iron and Steel Federation, *Monthly Report of the Iron and Steel Statistics,* for the years involved.
[a]Millions of tons of raw-steel equivalents.

bulk-cargo carriers, most of which were for export. Thus, a major portion of the steel consumed by the shipbuilding industry in Japan, and in some years this was as high as 6 million and 7 million tons, went into the export market.

The automotive industry was a creature of the post-war years. Before World War II, it was virtually nonexistent as production in 1938 amounted to 24,000 vehicles. In 1951, some 38,000 vehicles were produced. This was soon to change, however, and by 1963, output passed the 1-million-vehicle mark. By 1970, it was 5.3 million vehicles, and in 1980, the Japanese automobile industry led the world in production with 11 million vehicles. The automotive industry, like the steel industry, directed a very large percentage of its production to the export market. In 1970, it sold 1.1 million vehicles abroad. This figure grew constantly until in 1980 exports amounted to 6 million vehicles, or 55 percent of production. That year, steel shipments to the automotive industry were some 9.5 million tons, representing about 11 million tons of raw-steel equivalent, 55 percent of which was designated for export.[19]

These are but two examples of Japanese steel-consuming industries that direct much of their production to foreign markets. In total, it is conservatively estimated that one-half of Japanese steel production is destined for export, either as steel products or in the form of products containing steel.

The problem that faces the Japanese is whether or not they can continue to export large tonnages of steel into the world market, particularly in times of slack economic activity. In examining the Japanese exports by countries of destination, it is interesting to note that the largest tonnage, 6.2 million in 1979, was shipped to the United States, which has been for years either the first or the second largest steel producer in the world. The second largest market for Japanese steel is China, which took its highest amount, 5.6 million tons, in 1978, after which there was a significant decline.

The Japanese have sought to penetrate the Western European market, which they did quite successfully in the early 1970s by shipping a total of 2 million tons to that area in 1971. Here again, the countries of destination were large steel producers. Thus, Japan moved not only into markets such as the Third World where a steel deficit existed but also into those markets in the industrialized world where steel was a well-established industry.

In prosperous times, when steel demand was high and the domestic industries in Western Europe and North America could not fully supply their own needs, Japanese steel was welcome. However, in slack times or in depressed periods for the steel industry, when the industrialized countries are operating their industries at a relatively low rate of capacity, Japanese steel took a significant market share and was not welcome by local steel producers. Further, as the Third World develops its own steel industries, its need for Japanese steel will decline. Thus, this market, which in the 1960s and early 1970s was a rapidly growing one, will be less available to Japan in the future.

The industrialized countries in times of steel depression have set up barriers against Japanese steel. The EEC placed a quota of 1.2 million tons on Japanese imports, while Japanese steel to the U.S. market was restricted by voluntary quota agreements and, subsequently, by a trigger-price mechanism. Further, the Third World countries that have expanded their steel industries have cut back on Japanese imports to a greater or lesser degree. A prime example is Brazil, which imported 1.6 million tons of Japanese steel in 1974 when its own production was 7.5 million tons. By 1980, as the Brazilian steel industry increased its output to 15.3 million tons, or double the 1974 figure, imports fell to 160,000 tons, a decline of 90 percent.[20] This is an extreme case, but it does represent a trend.

Another Third World country that had been a large market for Japanese steel, having imported an average of 2 million tons a year from 1977 through 1980, is South Korea. In 1981, Pohang Steel Company of South Korea increased its capacity from 5.5 million tons to 8.5 million tons and installed a second wide hot-strip mill capable of producing 3 million tons of hot-rolled sheets. Since half of the Japanese exports to South Korea, or approximately 1 million tons, is in the form of sheet product, now that the Koreans apparently will be able to satisfy their own demand and these imports will be severely cut back. Further, the Koreans now have a capacity in excess of their own immediate needs and have not only become a competitor to the Japanese industry in the export market but have shipped steel to Japan.

Another country that furnishes the Japanese with a sizeable market for export is the Philippines. This will diminish, for plans are being made, with the help of the Japanese, to build an integrated mill in the Philippines that will provide 750,000 tons of steel, much of which will be turned into steel products that are now imported from Japan.

An example of another country that currently provides a market for Japanese steel but that could become a competitor is Taiwan. With the technical assistance of United States Steel Corporation, Taiwan has built a fully integrated mill that will produce plates as well as hot- and cold-rolled sheets. In 1979, Japan exported 1.6 million tons of steel to Taiwan, much of which was in the form of hot-rolled coils. Within a short time, probably by the end of 1982 or the beginning of 1983, Taiwan will have a hot-strip mill capable of producing more than 1 million tons of coil, as well as a cold-strip mill with a 700,000-ton annual capacity. Thus, the Japanese exports to Taiwan of approximately 600,000 tons of hot-rolled sheets and 375,000 tons of cold-rolled sheets will be in jeopardy as the new mills come into operation. They will be able to supply the entire domestic market, which is now dependent on Japan, and also will provide a surplus of approximately 400,000–500,000 tons for export in competition with the Japanese.

A further concern of the Japanese steel industry is the steel exports of other newly industrializing countries including Korea, Taiwan, India, Australia, Canada, Mexico, Brazil, Argentina, Spain, and South Africa. Exports from these countries have increased from a total of 7.6 million tons in 1973 to 18 million tons in 1979, and there is every indication that they will continue to grow. Table 4–8 indicates the size of this growth during the seven-year period from 1973 to 1979 inclusive.

There is little doubt that this development will have an adverse effect on Japanese exports throughout the world and especially in the Far East, which has heretofore been considered safe territory for the Japanese. In part, the Japanese are responsible for this development, since they have undertaken vigorous programs to sell steel-mill equipment throughout the world and particularly in the Far East. They are currently advising the Philippines, Thailand, and Malaysia on the installation of new steel mills through feasibility studies and the promotion of technical assistance, some of which involves the sale of steel-mill equipment. This problem is recognized, and efforts have been made to solve it by a number of means including joint ventures in the Third World countries. Brazil and the Far East provide examples of such joint-venture efforts.

Another step taken is the concerted effort by the industry to operate profitably at 70 percent of its capacity. Success has been achieved in this respect, particularly in 1979 and 1980 when, despite a 75 percent operating rate, the major Japanese steel companies were able to increase their profits significantly.

The Japanese currently have excess capacity in relation to world demand and this condition is likely to continue for a few years. A significant portion of the Third World capacity that is being installed is in the Far East and, consequently, represents competition for the Japanese, not only in the loss of the markets in those countries installing the capacity but also in those markets currently served by the Japanese to which the newly developed

Table 4-8
Steel Exports of Newly Industrialized Countries: 1973–1979
(millions of metric tons)

Country	1973	1974	1975	1976	1977	1978	1979
Korea	962	1,329	1,002	1,399	1,389	1,822	3,000
Taiwan	249	194	238	280	312	892	1,038
India	150	186	348	1,522	1,312	500	200
Australia	1,396	1,238	1,727	2,224	2,482	2,573	2,000
Canada	1,273	1,469	1,051	1,518	1,717	2,738	2,827
Mexico	141	122	76	154	275	350	400
Brazil	431	236	149	262	364	936	1,385
Argentina	625	380	54	356	280	743	450
Spain	1,712	800	1,580	2,443	2,678	4,117	4,100
South Africa	632	612	324	1,085	2,096	2,215	2,600
Total	7,571	6,574	6,549	11,243	12,095	16,886	18,000

Source: U.N., Study by Japanese Steel Export Manufacturing Committee.

countries will attempt to ship steel. Should world demand increase by approximately 1.5 percent a year for the next five years, it will absorb all of the new capacity projected by the Third World and also will permit the Japanese to take advantage of this growth in demand by drawing on their currently idle capacity, which can be put into operation very easily. However, when the demand for steel is sluggish in the years ahead, Japanese steel exports will be affected adversely. To counter this, the Japanese have vigorously promoted the sale of products made of steel, including structures and manufacturing equipment, that provide them with additional outlets for their steel-mill products as well as steel in fabricated form.

Ownership

The Japanese industry before World War II was quite small, with an annual production in the early 1930s of less than 3.0 million tons. A fragmented industry, it underwent structural change in 1934 when several firms were combined to form the Japanese Iron and Steel Company Limited. The principal operation was at the Yawata state-owned plant—to this were added six private steelmakers—and the resultant entity was described as a semigovernment, semiprivate organization that increased in size up to World War II. At the end of the war, with the reconstruction of much of the Japanese economy, Japan Iron and Steel Company was split into two firms: Yawata Steel and Fuji Steel, which were private companies.

Other steel companies in existence at that time included Nippon Kokan, an integrated operation, and Kawasaki Steel, Sumitomo Steel, and Kobe Steel, which were nonintegrated. These last three gradually expanded and integrated their operations. A seventh company, Nisshin Steel, was formed in 1959 by a merger of two smaller companies. It has grown considerably since that time and is fully integrated. Thus, today all of the fully integrated steel companies are privately owned.

In 1970 a merger was effected between Yawata and Fuji, the two largest producers in Japan. The resultant entity, Nippon Steel, became and has remained the largest steel company in the world.

In addition to the integrated steel producers in Japan, a multitude of electric-furnace steelmaking companies exist that vary in size from capacities of 2.0 million tons to those with approximately 50,000 ton capacity. The vast majority of these are privately owned.

Future Developments

The Japanese steelmakers are much more optimistic about the future than most of their counterparts in the industrialized world; for, although they have no immediate plans for growth, they are convinced that steel has a bright future and, in keeping with this, they are improving an already efficient industry. While some steelmakers in other parts of the world talk of the decline in steel and the inroads made on the steel market by other materials, the Japanese, in a report issued in mid-1982, state:

> The steel industry will remain the principal provider of basic materials for all industries. Both in the near and distant future, there will emerge no viable alternative to steel (in availability, price and quality).[21]

Further, in assessing the significance of steel to Japan, the report states:

> Every step taken by the steel industry in its progress is a mirror image of the progress of the Japanese economy itself.[22]

The Japanese point with pride to world records established by their steel plants and are willing to offer technical cooperation anywhere in the world. In 1981, the steelmakers in Japan signed fifty-eight contracts with forty companies in twenty-two countries. The Japanese recognize the assistance given to them by the United States and some Western European countries in rehabilitating their steel industry after World War II and they feel that they should return the favor.

In terms of international trade, since the Japanese are the leaders in steel exports, they are very much in favor of orderly marketing and claim to have set an example by exporting steel in an orderly fashion "as part of the national policy to maintain the free-trade system."

Programs for continual improvement of the Japanese steel industry involve ongoing investment in steel plant and equipment directed at minimizing costs, developing new products with a higher value added, and diversifying business operations into engineering and chemicals.

One of the prime targets in the future operation of the steel industry is the reduction in the use of energy.

Further, there is a comprehensive program aimed at restructuring the electric-furnace segment of the steel industry. It is aimed at eliminating excess capacity as well as the "excessive competition characteristic of this industry." So that this program can be carried out, the member companies are asked to contribute to a fund that will facilitate the retirement of equipment.

With its plans for continual modernization, the Japanese steel industry throughout the 1980s will remain the highly competitive force that it has been in the past.

Notes

1. IISI, *World Steel Figures 1981* (Brussels), p. 8.
2. Japan Iron and Steel Federation, *Statistical Yearbook 1972,* p. 3;

and Japan Iron Federation, *Monthly Report of the Iron and Steel Statistics* (March 1981):3.

3. IISI, "Report of Proceedings of the Seventh Annual Conference," Johannesburg (Brussels 1974), E,716,4.

4. IISI, "Report of Proceedings of the Seventh Annual Conference," Johannesburg (Brussels 1974), E,716.

5. IISI, "Report of Proceedings of the Seventh Annual Conference," Johannesburg (Brussels 1974), E,716,7.

6. Japan Federation, *Statistical Yearbook, 1972,* p. 4; and Japan Federation, *Monthly Report of the Iron and Steel Statistics* (March 1981):3.

7. Conversations with Nippon Steel executive personnel.

8. Conversation with the president of Kawasaki Steel.

9. Japan Federation, *Monthly Report of the Iron and Steel Statistics* (March 1981):20.

10. Japan Federation, *Statistical Yearbook 1972,* p. 4.

11. Ibid.

12. Japan Iron and Steel Federation, *The Steel Industry of Japan 1973,* p. 13.

13. Japan Federation, *Statistical Yearbook 1972,* p. 4; and Japan Federation, *Monthly Report of the Iron and Steel Statistics* (March 1981):32.

14. Japan Federation, *Monthly Report of the Iron and Steel Statistics* (March 1981):32.

15. The prices of bunker oil vary somewhat in different sections of the world.

16. Japan Federation, *Statistical Yearbook 1972,* p. 4; and Japan Federation, *Monthly Report of the Iron and Steel Statistics* (March 1981):3.

17. *World Survey of Continuous Casting Machines for Steel,* 6th ed. (Zurich, Switzerland: Concast Documentation Center, 1980), pp. 70–88.

18. Japan Federation, *Statistical Yearbook 1972,* p. 3.

19. Japanese Automobile Manufacturers Association, Inc., *Motor Vehicle Statistics of Japan 1981* (Tokyo, 1981), pp. 10 and 17.

20. Japan Federation, *Monthly Report of the Iron and Steel Statistics* (March 1981):17.

21. Nippon Steel Corporation: *Current State and Problems of the Japanese Steel Industry* (Tokyo: June 1982), p. 3.

22. Ibid.

5 The United States Faces Critical Decisions

The U.S. steel industry fared much better than its Western European counterparts in the second half of the 1970s. Four out of five years were profitable for the industry as a whole, although individual companies incurred losses and some were forced into bankruptcy. The one profitless year was 1977, when there was a sharp increase in low-priced imports and a number of plants were shut down.

The industry currently is composed of fourteen fully integrated companies, ranging in size from United States Steel Corporation, which produced 30 million tons of raw steel in 1979, to relatively small companies with a production of less than 2 million tons in that year. In addition to the integrated enterprises, over sixty companies make steel in electric furnaces.

Many changes have taken place in the past fifteen years that have significantly reduced the number of integrated steel companies. In 1968, there were twenty-three independent, integrated steelmakers, but through mergers, changes in corporate structure, and bankruptcies, this number has almost been cut in half. Alan Wood Steel and Wisconsin Steel have shut down completely, while Granite City Steel, Youngstown Sheet and Tube, Detroit Steel, and Wheeling Steel left the ranks of the independents through mergers. Crucible Steel and Cyclops discontinued blast-furnace operations, and Kaiser Steel is dropping its coke-oven, blast-furnace, and steelmaking facilities. Sharon and Jones & Laughlin are still operated as integrated plants but ceased to be independent when they became part of conglomerates. The remaining integrated companies are:

United States Steel Corporation,

Bethlehem Steel Corporation,

Jones & Laughlin Steel Corporation,

Republic Steel Corporation,

National Steel Corporation,

Armco Steel Corporation,

Inland Steel Company,

Wheeling-Pittsburgh Steel Corporation,

Sharon Steel Corporation,

McLouth Steel Corporation,

Interlake Inc.,

Steel Division of Ford Motor Company (Rouge Steel Company),

CF&I Steel Corporation,

Lone Star Steel Corporation.

Market forces in the United States have restructured the U.S. steel industry and reduced its size. The same objective should be achieved in the EEC countries of Western Europe by the Davignon plan.

The electric-furnace producers range in size from Northwestern Steel and Wire, which is.capable of producing 2.5 million to 3.0 million tons annually, to small minimills with capacities of 100,000 tons or less.

Growth

Between 1950 and 1970, the steel industry in the United States experienced a period of substantial growth. In 1950, capacity was 90 million tons, with a production in that year of 88 million tons, a virtual full-capacity operation. During the 1950s, partly under government pressure and partly because of the increased demand for steel, first due to the Korean War and subsequently because of the rapid growth of the economy, the steel industry added 46 million tons, so that by 1960, capacity reached 136 million tons, a 50 percent increase. The 46 million tons of additional capacity were installed evenly through the decade, with 24 million tons in the first five years and 22 million tons in the last five years. During the 1960s, very little was added to steelmaking capacity; however, a great deal of capital was invested in replacing obsolete equipment with modern facilities. By 1970, capacity had grown to approximately 144 million tons, a relatively small increase during the ten-year period.

The decade of the 1950s was one of growth, while the 1960s was a period of modernization. The 1970s began on a low note, but in 1973 and 1974, the United States participated in a worldwide boom and established new production records as output reached 137 million tons in the first year and declined somewhat to 132 million tons in the second.

During this prosperous period, plans were made by most of the steel companies to expand capacity to meet projected future growth. Aggregate expansion planned by the industry totaled 21 million tons; in addition, substantial facility replacements of more than 20 million tons of raw-steel capacity were announced bringing the total amount of steelmaking capacity planned for installation between 1973 and 1980 to approximately 41 million tons. This melting capacity was to be augmented by a number of new finishing facilities.

Two fully integrated greenfield-site plants were projected: one by the United States Steel Corporation at Conneaut, Ohio, and the other by National Steel Corporation at its Midwest location near Chicago. Most of the additional 21 million tons were to be obtained through expanding and rounding out existing facilities. Some of the major installations included blast furnaces, basic-oxygen steelmaking facilities, large electric furnaces, plate mills, pipe mills, bar mills, and other finishing facilities. A total of twenty-nine companies, including minimills as well as integrated producers, formulated expansion plans.

When demand dropped sharply in 1975 and did not recover strongly in 1976, a number of these projects were either postponed or canceled, resulting in the installation of only a small part of the projected 21-million-ton growth, most of which was to come through rounding out existing facilities. How successful this was is still a matter of conjecture because the industry has not been pushed to the limit since 1974 and because there is no precise method of judging whether the roundout, which in a number of instances consisted of breaking bottlenecks, achieved the increased tonnage that was hoped for.

The increase that took place came from new facilities such as Inland's large blast furnace and electric furnaces added by United States Steel, the Ford Motor Corporation Steel Division, and Northwestern Steel & Wire, as well as some new and expanded minimills. This equipment was installed at existing plants, because, with a downward revision of future demand, as well as increasing facility costs, one of the two greenfield-site plants was canceled and the other indefinitely postponed.

As economic conditions in the steel industry worsened in 1977, a number of companies made plans to reduce capacity by closing obsolete facilities. In the four-year period from 1977 to 1981, some 12.5 million tons of steelmaking capacity were withdrawn from production with no plans for replacement. The withdrawals consisted of substantial tonnages that were lost when Bethlehem Steel, United States Steel, Youngstown Sheet and Tube, National Steel, Crucible, and Cyclops shut down plants or parts of plants and when Alan Wood and Wisconsin Steel went into bankruptcy.

As a consequence, the U.S. steel industry actually had less capacity in 1981 than it had in 1974. In the latter year, the industry had a capability of producing approximately 145 million tons of raw steel, and that figure fell to less than 136 million tons by 1981.

The only recent expansion in steel-industry capacity has been in the minimill sector. Between 1977 and 1982, the minimills have added about 4.5 million tons through the construction of five new plants and the expansion of a number of existing facilities. All but one of the new plants were built by existing minimill companies that expanded their activities at new locations.

Despite the fact that the integrated producers are not adding new steel-making facilities, their output of finished steel will be increased in the next two to three years through the installation of more continuous-casting equipment. In 1980, the U.S. industry cast some 20.6 million tons of steel, or 20 percent of its production. With current plans for additional casters, it is quite conceivable that this production could be doubled by 1985 or 1986. Such a move would increase yield and provide at least 2 million tons more of steel shipments.

Currently, not only are there no plans for expanding capacity among the integrated producers, but also it is more than possible that some of the present steelmaking facilities will be either mothballed indefinitely or closed. An example is the recent announcement by National Steel that it would reduce the capacity of its Great Lakes plant from 5 million tons to 3 million by shutting down two blast furnaces. Kaiser Steel wrote off 3 million tons of steelmaking capacity, and several other companies, with marginal facilities, are in a somewhat precarious position and might not be able to withstand a prolonged steel recession. Further, a number of open-hearth shops will be shut down, some of which will not be replaced. These actions could reduce capacity by another 3 million to 4 million tons, neutralizing the additions made by the minimills. It should be noted that mini-mill growth is confined to a limited number of products, principally bars and some rods, while the elimination of integrated capacity affects a much broader product range, including sheets and plates.

The U.S. steel industry, during the first half of the 1980s, will concentrate on modernizing its basic steel-producing and finishing equipment. Some new facilities will be added, like seamless-tube mills, several of which have already been announced by CF&I, United States Steel and some new companies. Further, continuous casting will have high priority in terms of new installations. Several large units are already under construction, such as those at Northwestern Steel & Wire, Republic Steel at Cleveland, United States Steel at Lorain, Ohio, and Bethlehem Steel at Sparrows Point. In 1980 and 1981, three continuous-casting units were installed and twelve others were either on order or in the process of construction.

To complete all the investment plans announced by the industry, huge sums of money will be expended in the years ahead, averaging $3 billion or more annually. Some indication of the need for these sums can be had from a brief examination of the costs of building facilities. A 5,000-ton-a-day blast furnace was planned by one of the companies. However, when the price reached $300 million, it was indefinitely postponed. A continuous hot-strip mill, which was installed in the late 1960s for $125 million to $150 million, would involve an investment in the early 1980s in excess of $350 million. A 300,000 ton-a-year seamless-pipe mill requires an expenditure in the area of $300 million.

Most steel companies feel that the demand for steel will increase during the current decade, but opinions differ as to the amount. Some companies place the rate of growth at a very low figure—less than 1 percent—and feel no need to expand capacity. Others are more optimistic and would like to increase their facilities, yet the huge capital sums required for additional plants and equipment, over and above replacement investment, have dampened the ardor for expansion. The combination of high capital-investment costs, some uncertainty about the rate-of-demand increase, and the share of the market taken by imports has eliminated the possibility of constructing a greenfield, fully integrated steel mill for the next five years and most likely for the remainder of the decade. This is reminiscent of decisions made by four integrated producers in the 1960s, when some of the same reasons prompted them to cancel plans to build fully integrated plants on the West Coast of the United States. United States Steel, Bethlehem, Armco, and National Steel all had extensive landholdings in the San Francisco area on which they had contemplated the construction of integrated steelworks.

In terms of new facilities, primarily because of cost, it is virtually certain that, with one possible exception, no new blast furnaces will be built, either for expansion or replacement, until after 1985. The one possibility is at the Middletown plant of Armco Steel, where a new blast furnace might be built to replace two old furnaces located a few miles distant at the Hamilton Works. The existing units in other companies will be revamped and improved but not replaced.

As a consequence of these decisions, it is most probable that raw-steel capacity in the United States will be held below 130 million tons. Should steel demand require the production of 127 million tons or more of raw steel during the next few years, capacity will be pushed almost to the limit and will result in a tight supply situation. This amount of output is certainly possible in the years ahead, since production in 1978 and 1979 reached 124 million tons and 123 million tons respectively. Production in 1980 fell sharply to 102 million tons, as a result of an unusual dip in the economy in the second half of that year, and recovered in 1981 to 109 million tons.[1] A recovery in economic activity in the years ahead could easily restore steel production to the 1978–1979 levels, and with a capacity of less than 130 million tons, this would mean an operating rate of over 90 percent.

If the industry were to modernize its plant and equipment so that virtually all of its facilities are efficient, it would operate at these high levels and earn profits on every ton produced. Heretofore, when operating rates passed 90 percent, marginal and, in some cases, obsolete facilities were pressed into service without producing profits.

As of 1981, a plant-by-plant survey of the U.S. steel industry indicates that between 70 and 75 percent of the basic production facilities and finishing equipment are modern and fully competitive, while 25 to 30 percent are in need of replacement.

The survey revealed that all of the facilities involved in oxygen steelmaking are twenty years old or less, many of them are less than ten years old, and virtually all are fully competitive. Basic-oxygen steel constituted 61 percent of the industry's output in 1981. Electric-furnace capacity is the largest of any steel industry in the world, and three-quarters of it has been installed since 1960. One-half of the total has been installed since 1970 and is fully modern and competitive; in fact, electric-furnace units in the United States are superior to most in the world. The hot-strip mill is the most significant rolling facility in the steel industry, since approximately one-half of the steel processed goes through it. The U.S. industry installed twelve new hot-strip mills between 1961 and 1974, and they are the equal of any in the world and superior to many. Examples of other modern equipment could be multiplied, but these will suffice to indicate the state of the industry.

The U.S. steel industry must modernize fully with the latest technology in the years ahead if it is to regain the competitive position in the world that it held at the close of World War II. It must continue to subscribe to the change in philosophy that has taken place in the last few years, whereby the emphasis is no longer placed on tonnage but on efficient operation and profitable tonnage. As a result of this new approach, some unprofitable products have already been dropped and more will follow.

Profits

In terms of profitability, it should be noted that during the last half of the 1970s, the U.S. industry, although faced with a number of difficulties, including a low return on sales and a lack of adequate cash to modernize as completely as it would like, was still profitable in relation to other steel industries in the world. The contrast is particularly striking when compared with the European industry, which has suffered very heavy losses.

The high point in dollar earnings was reached in 1974, as the industry earned $2.5 billion. With the decline in production and shipments in 1975, profits fell to $1.6 billion; this slide continued through 1976 as profits dipped to $1.3 billion.[2] It is interesting to note that 1976 shipments were some 9 million tons above 1975 shipments, yet profits fell. This was due in great part to increased costs and a weakness in prices brought on by low import prices that forced the U.S. prices down.

In 1977, earnings on an industrywide basis disappeared. The American Iron and Steel Institute indicated earnings of $22.3 million in that year on sales of $39.4 billion, or a small fraction of 1 percent.[3] A number of companies sustained severe losses and one went bankrupt. Bethlehem Steel posted a very large loss due to the impact of plant closings. However, much of the reduction in earnings was due to severe competition from imports that came into the country at prices as much as $100 per ton below the U.S. list price. This made a shambles of the price structure, which was reflected in the lack of earnings.

In 1978 and 1979, with a pickup in the economy and the institution of the trigger-price mechanism against import dumping, profitability returned. The industry posted earnings in those years of $1.3 billion and $1.2 billion respectively. In 1980, net income rose to $1.6 billion, despite a drop in shipments of 15 million tons.[4] This was the reverse of the situation in 1976 when profits declined as shipments rose. The return on sales for 1980 amounted to 2.9 percent, hardly a satisfactory figure, although envied by the Western European producers who were mired in deep losses.

In 1981, the industry registered a $2.4 billion profit, second only in dollar terms to 1974. Somewhat more than half of this profit, $1.365 billion, was earned in the first six months when the operating rate rose as high as 89 percent at one point. In the third quarter, earnings were in excess of $1 billion, bringing the total to $2.437 billion. It should be pointed out that a significant portion of this came from the sale of assets such as raw-material deposits of coal. In the fourth quarter, with a sharp falloff in orders and operations, there was an actual industry-wide loss of some $34 million.

The first quarter of 1982 witnessed a severe drop in steel shipments to 17.2 million tons, one of the lowest quarterly rates in the postwar period. This drop resulted in large losses for most of the companies, while those that succeeded in avoiding losses made very slim profits. The remainder of 1982 looks bleak indeed, and most companies project further losses.

Despite the earnings in 1981, the losses in 1982 will make it difficult for the companies to cover needs for full modernization of plant and equipment. One of the reasons for this is the amount of money still required for pollution controls for both air and water. Between 1968 and 1980, total funds spent for this purpose amounted to $4.4 billion.[5] This was a necessary expense to improve the environment and can be regarded as a social cost of doing business.

The steel companies in the United States and, for that matter, throughout the world must experience at least three consecutive good years in terms of operating rates and profitability and be convinced of the growth in long-range steel demand before any additional capacity is contemplated.

Raw Materials

The U.S. steel industry was particularly fortunate in respect to raw materials from its beginnings in the latter part of the nineteenth century through the close of World War II. Rich iron-ore deposits were found in the Lake

Superior region, and with the rate of use before World War II, they were calculated to last to the end of the twentieth century. Coking coal of the highest quality was available in large tonnages throughout western Pennsylvania, West Virginia, Kentucky, and Tennessee.

Ore

As a result of World War II, almost a half billion tons of the rich Mesabi Range ore were consumed, and with the postwar increase in demand and production levels, it was virtually certain that ore consumption would continue at least at the wartime rate. Under these circumstances, estimates were made in 1947 that the Mesabi Range, the principal source of U.S. ore, would be depleted within twenty-five years and, therefore, it was imperative that additional ore be found.

Two sources were possible: (1) new and, hopefully, rich iron-ore deposits might be discovered outside of the United States, and (2) the lean ores, which existed in abundance in the United States, could be beneficiated to a point where they would be acceptable blast-furnace feed.

A remarkably successful, worldwide search for iron ore was undertaken. Extensive high-grade deposits were found by U.S. companies in Venezuela, West Africa, and Canada in the late 1940s and early 1950s. In the 1960s, other large, rich deposits were discovered in Brazil and Australia that added enough ore to the world's reserves to care for the steel industry's needs for a century.

In addition to the search for deposits, a substantial investment was made in beneficiating equipment, particularly pelletizing facilities to crush, concentrate, and upgrade lean ores in the Lake Superior region. Thus, the United States solved its ore-supply problem in a twofold manner. At present, it imports high-grade ores from Brazil and Venezuela, as well as upgraded ores from Canada. Further, pellet plants, constructed in the United States and Canada, provide a capacity to produce about 120 million tons of product from beneficiated lean ores. Table 5-1 lists the pellet plants in the United States and Canada with their capacities. Canada is included because many U.S. steel companies have interests in and obtain pellets from Canadian pellet plants.

Substantial investments have been made in iron-ore properties outside of the United States by a number of U.S. steel companies. To recoup these, the industry imports from several countries between one-fourth and one-third of the ore required in the United States. Table 5-2 gives ore imports into the United States by principal countries of origin, as well as total consumption and the percentage represented by imports.

Between 1970 and 1976, Canada's share of U.S. iron-ore imports fluc-

Table 5-1
North American Pelletizing Capacity
(millions of gross tons)

Plant	Design Capacity	Plant	Design Capacity
United States		Black River Falls	0.9
		Eagle Mountain	2.5
Tilden	8.0	Atlantic City	1.6
Empire	8.0	Fire Lake	6.0
Marquette	3.5		
Butler	2.7		
National	5.8	Canada	
Groveland	2.0	Sherman	1.1
Erie	10.6	Adams	1.1
Hibtac	8.1	Iron Ore Company	
Eveleth	5.9	of Canada	16.0
Minntac	18.0	Wabush	6.0
Minorca	2.6	Griffith	1.5
Reserve	9.5	Total	121.4

Source: Cleveland-Cliffs Iron Company.

tuated from 40 to 50 percent. Since 1976, it has risen significantly, running as high as 67 to 68 percent in some years. This is not surprising in view of the fact that a number of U.S. steel companies have large investments in Canadian ore bodies on a take-or-pay contract basis. Two of the principal companies in Canada with U.S. participation are the Iron Ore Company of Canada and Wabush Mines. The steel companies in the United States participating in the Iron Ore Company of Canada include Bethlehem Steel, National Steel, Republic Steel, Armco, Jones & Laughlin, and Wheeling-Pittsburgh. In Wabush Mines, the participants from the United States include Jones & Laughlin, Inland Steel, Wheeling-Pittsburgh, and Interlake. In addition, the United States Steel Corporation operates a subsidiary, Quebec-Cartier Mines, and has a participation in the the Fire Lake Project.

Venezuelan imports have declined since the nationalization of iron-ore production in 1975, eliminating the United States Steel Corporation and Bethlehem Steel as operating companies.

In the years ahead, the U.S. industry will continue to depend on imports, particularly from Canada, for a part of its requirements. The major portion of the total requirement will be satisfied by domestically produced pellets formed from ore bodies in the Lake Superior region, as well as other smaller deposits in a variety of locations throughout the country, such as

Table 5-2
U.S. Iron-Ore Consumption and Imports, by Principal Countries of Origin: 1970–1980
(millions of gross tons)

Year	Total Consumption	Total Imports	Import Tonnage by Major Country of Origin					Percentage: Imports of Consumption
			Canada	Venezuela	Liberia	Brazil		
1970	144.6	44.9	23.9	13.0	1.9	2.0		31.1
1971	129.7	40.1	20.3	13.0	1.8	1.8		30.9
1972	143.8	35.8	18.2	10.9	2.8	1.1		24.9
1973	162.2	43.3	21.6	13.1	2.7	3.2		26.7
1974	155.2	48.0	19.7	15.4	2.7	6.6		30.9
1975	126.1	46.7	19.1	13.1	2.5	7.5		37.0
1976	134.5	44.4	25.0	9.0	2.2	5.4		33.0
1977	126.4	37.9	25.3	6.2	1.8	2.2		30.0
1978	134.3	33.6	19.2	6.1	2.2	4.0		25.0
1979	134.0	33.8	22.6	4.6	2.2	3.1		25.2
1980	105.9	25.1	17.3	3.6	1.6	2.0		23.7

Source: American Iron and Steel Institute, *Annual Statistical Reports*, various years.

New York, eastern Pennsylvania, Alabama, California, and Wyoming. The taconite deposits in the Lake Superior region are extensive and will last for many decades, but the ore in the other areas is limited.

Virtually all of the ore charged into the nation's blast furnaces is processed into pellets or sinter. In 1980, of the 105.8 million tons consumed, 100.3 million were in the form of pellets or sinter, with the balance very heavily in favor of pellets by more than two to one. Thus, iron ore, as it is used today, is no longer, strictly speaking, a raw material. It is processed in plants with equipment requiring capital investments running into many millions of dollars.

Pellet plants constructed in the last three years required an investment of $90 per annual ton of capacity, as compared with $51 in the early 1970s.[6] A new sinter plant, projected for construction in 1982, will require about $35–$40 per ton of annual capacity. Pellet and sinter plants previously constructed were far less expensive and are still in use.

Coal

The United States has very large reserves of metallurgical coal needed to produce coke for its blast furnaces. Not only are the reserves adequate for the needs of the domestic industry for decades to come, but also they are extensive enough to permit substantial exports to many of the steel-producing countries throughout the world. Recent surveys have identified premium-grade coking coals in the United States and estimated their tonnage at 19.6 billion tons. These coals are in three classifications: high, medium, and low volatile. Low-volatile coal contains 14–21 percent volatile matter, medium-volatile coal contains 22–28 percent, and high-volatile coal has over 28 percent. The high-volatile coal represents the largest reserve (13.4 billion tons), while the medium- and low-volatile-coal reserves, although substantial, are by no means as great (3 and 3.1 billion tons respectively).

Medium-volatile coal is usually the most desirable charge for the coke ovens and is achieved in many instances by a blend of high- and low-volatile coals. Most of the metallurgical coal reserves are located in West Virginia, which accounts for 10.5 billion tons of the total. Kentucky has 5 billion tons, and Virginia has 1.6 billion tons. There are significant reserves in a number of other states; however, in relation to the aforementioned three, they are relatively small.

Since the end of World War II, coal exports have constituted a significant part of the industry's business as the U.S. coal companies exported huge tonnages of metallurgical coal. In 1979, the figure was 50.7 million tons, and this increased to 63.1 million tons in 1980. The principal recipients in 1980 were Japan, with 21.9 million tons; the EEC, 20.5 million tons;

Canada, 6.3 million tons; and South America, 4.2 million tons, where
Brazil accounted for 3.2 million tons.[7]

Although the major producers in the coal industry are coal companies,
the steel companies that have blast furnaces and coke ovens, and consequently a need for coking coal, have either their own coal mines or substantial interests in coal mines. Further, several foreign steel companies have
invested in coal-mining operations in the United States, including the
French, Italians, and ARBED, the Luxembourg-based steel company.

Coal production for use in the steel industry's coke ovens during the
past twenty years has ranged from a high of 87.2 million tons in 1970 to a
low of 69.4 million tons in 1979, all of which was produced in the United
States. It should be noted that the United States does import some bituminous coal; in 1980, the figure was 895,000 tons, of which 770,000 tons, or 86
percent, came from South Africa. By contrast, U.S. production of bituminous coal was in excess of 800 million tons.

The statistics on coal reserves indicate that no problem exists with the
future supply of coking coal for the U.S. industry. With reserves in excess
of 19 billion tons and an annual production of less than 100 million tons,
the coking-coal supply will be adequate for the indefinite future.

Although no problem exists about the future supply of coal, the U.S.
steel industry faces some difficulties in the production of coke. The ability
of the industry to produce coke has been declining over the last three decades. In 1950, production was 66 million tons; in 1973, a year of record
steel output, coke production stood at 58.3 million tons; and in 1978, it fell
to 44.1 million tons. In this last year, the United States was forced to import
5.2 million tons to meet its blast-furnace requirements.[8] It should be noted
that some of the decline in coke production is a result of current blast-furnace practice that has improved to a point at which less coke is required per
ton of iron produced. In 1956, the U.S. industry consumed 0.78 tons of
coke for every ton of pig iron produced; by 1980, this had fallen to 0.5 tons,
a decided improvement. In the past year, this has shown signs of further
improvement as coke rates for a number of blast furnaces have been reduced by as much as 91 kilograms per ton principally due to the production
of higher-stability coke.

Coke ovens in the United States are aging and must be replaced, since
the older units are less productive. About 30 percent of capacity is over
twenty-five years old. Further, the Environmental Protection Agency, during the 1970s, recommended the closure of many coke-oven batteries to
meet air-pollution requirements. An extension of the original closure date,
which was scheduled for December 31, 1982, has been given by the Steel
Industry Compliance Extension Act of 1981. The deadline in now December 31, 1985, so that more time is available to replace the aging ovens. The
capital investment required has increased significantly during the last ten

years to a point at which, in 1981, the cost of installing one annual ton of capacity for a by-product coke plant on a greenfield site was in excess of $250, as compared with $75–$85 in 1970.[9] Despite the high cost, a number of coke ovens will be replaced, but it is questionable as to whether capacity will be adequate to meet the demands of the steel industry operating at 85 percent or higher for an extended period of time. It is more than likely that coke will be imported to meet the requirements of the U.S. steel industry at a sustained level of operations above 85 percent.

During the 1980s, metallurgical coal will be plentiful, but coke could be in short supply from time to time, depending on the rate of steel-industry operations.

Scrap

Scrap is a vital material for the steel industry in the United States in view of the large output of the electric furnace. In 1979, electric-furnace production reached an all-time high of 30.8 million tons, a fourfold increase over the 1960 level of 7.6 million tons.[10] With approximately a 10 percent loss in yield from the charge to the molten steel, scrap requirements for the electric furnaces in 1979 were in the area of 35 million tons.

A small production of DRI, less than 1 million tons, was also used as charge material for some electric furnaces. As indicated, the development of direct-reduced material has a number of hurdles to overcome before it can be made available adequately in the United States. As a result, scrap will continue to be the overwhelmingly dominant electric-furnace feed material for the remainder of the decade.

It is expected, from a review of installation plans, that electric-furnace capacity will be approximately 41 million tons by 1985 and could reach 48 million tons by 1990. If the furnaces operate at near full capacity in 1985, they would require some 45 million tons of scrap. Further, since many of the units recently installed are intended to produce high-quality steel to be used for plates, pipe, sheets, and tin-plate, they will require high-grade scrap with a minimum of contaminants or tramp elements. Unfortunately, as requirements for electric-furnace steel are increasing, the quality of scrap is declining.

Scrap comes from three sources: (1) mill revert that is generated within the plant itself as the raw steel goes through a number of processes to the finished product, (2) prompt industrial that results from crops and trimmings generated by companies making products from steel, and (3) obsolete metal that comes from discarded steel products.

Revert scrap is considered a thoroughly dependable material since its chemistry is known. This will continue for the most part to be a dependable source. However, with the installation of continuous-casting units, the amount of revert scrap will diminish significantly, since the yield from raw steel to finished product will be increased by approximately 10 percent, leaving that much less revert scrap. Further, a number of steel plants are making high-quality low-alloy steel with the result that the revert scrap contains some of these alloys. These have to be segregated, particularly when scrap is used for carbon-steel production. As yet, the change in revert scrap does not present a great problem, but it seems that it could increase in the years ahead.

Prompt industrial scrap has undergone a change in quality. The examples of this stem from recent developments in the automobile industry that provides a very large amount of prompt industrial scrap. The automobile manufacturers are using more zinc-coated material to protect against corrosion, and zinc is an undesirable element in scrap. Thus, this major source of prompt industrial scrap now must offer some material that contains a contaminant. In the years ahead, the automobile makers plan to use more and more galvanized steel in their bodies; in fact, some may shift to an all-galvanized body, and this move will certainly aggravate the problem.

Obsolete metal is currently presenting difficulties to some electric-furnace operators. They maintain that quality is deteriorating to a point at which the metal that is sold today as number-one grade is reminiscent of number-two grade in the 1960s and early 1970s. Several electric-furnace operators also complain that the use of plastics combined with steel in the production of many products makes it difficult to separate these materials in scrap preparation, thus introducing further contaminants.

Problems also exist in terms of geographical availability of scrap. If it is plentiful on the West Coast, this brings little comfort to the electric-furnace operators in the midwest, since the freight rate from California to Chicago is enough to make the use of that scrap uneconomical.

It should be pointed out that there is an adequate supply of scrap, both in terms of quality and quantity, for the less-sophisticated steel products like concrete reinforcing bars. However, these products will constitute a smaller share of electric-furnace steel production as the furnaces continue to turn out higher-quality steels.

The United States is an exporter of scrap and will continue to be during the remainder of the decade. However, as previously pointed out, the Japanese—the principal customer for the U.S. scrap industry—require less imported material since they are generating more at home. Other large consumers—Spain, Italy, and Mexico—should remain significant markets, as will Japan, even though it will require less. One balancing factor in Japan is the growth of continuous casting, which cut down on mill-revert scrap.

In the years ahead, when steel demand is high and production at peak levels, there will be problems in supplying adequate-quality scrap to meet the requirements of a part of the electric-furnace sector of the steel industry. This problem is recognized by a number of electric-furnace operators and has turned their attention to the possibility of obtaining DRI to enrich the electric-furnace charge. At present, this material is not abundantly available either in the United States or on a worldwide basis; however, improvements in its production are being made, and greater quantities will be turned out in the years ahead, making some tonnage available for export into the United States.

Technology

The adoption of new technology by the various steel industries throughout the world has been influenced by the circumstances in which each found itself during the 1950s and 1960s. The U.S. steel industry held a unique position at the close of World War II because, unlike the industries in most of the industrialized world that had been subject to extensive war damage, the U.S. industry was not only untouched but also added extensive new capacity.

Open Hearth

In 1945, the official steel capacity was 87 million tons. Of this amount, some 12.7 million tons, mostly open-hearth capacity, were added during the war, representing the latest steelmaking technology at that time.[11] The open-hearth furnace accounted for some 88 percent of total capacity, with the remainder divided almost equally between the Bessemer converter and the electric furnace. Of the total capacity, a considerable amount of both melting and finishing equipment was obsolete, yet it was kept in operation during the war because of the need to produce steel at any cost by any type of facility.

At the end of the war, several steel companies were anxious to close down obsolete facilities and replace them with modern, efficient units. In fact, some closures were definitely planned for 1946 and 1947, so that the actual capacity figure published by the industry was reduced to 82.7 million tons in 1947.[12]

In 1948, as steel demand rose to its peacetime peak, some of these condemned units were put back into service to meet the nation's steel requirements. At that time, a controversy arose as to whether or not there was enough steel capacity to care for the growing needs of the economy. Steel-industry representatives looked on the growing demand as a short-term, postwar boom that had developed to satisfy pent-up demand caused by wartime shortages in civilian products. Once this demand was taken care of, they argued, it would drop, and then the current steel capacity would be adequate. A number of these representatives remembered well what had happened after World War I when the immediate postwar boom was followed by a severe depression. Many expected history to repeat itself. This did not happen, principally because the backlog generated during the war was much greater than anticipated and, to a significant extent, was supplemented by other backlogs in certain areas of the economy like construction, which had been built up during the depression years of the 1930s.

In opposition to the industry's views, some government economists maintained that an increase of 10 percent in steelmaking capacity was needed to care for the growing economy. The dispute reached a climax in January 1949, when President Truman, in his State of the Union message, declared that he wished:

> [T]o authorize an immediate study of the adequacy of production facilities of materials in critically short supply, such as steel; and if found necessary, to authorize government loans for the expansion of production facilities to relieve such shortages, and furthermore to authorize the construction of such facilities directly if action by private industry fails to meet our needs.[13]

Plans were made by some companies to expand capacity, but when the Korean War broke out in mid-1950, not only were these accelerated but also the remaining companies announced expansions.

Between 1950 and 1953, seven new open-hearth shops with a total of forty-three furnaces and a capacity of nearly 20 million tons were installed. During those years, this was the only type of facility with a proven record that could produce the required tonnage. The technological breakthrough that used oxygen to make steel was not operative until 1952 when a small vessel with a 30-ton-per-heat capacity was built at Linz in Austria. This process was not considered a proven success until 1957, by which time the steel industry in the United States had installed over 35 million tons of new open-hearth capacity.

Years later, government and academic economists criticized the industry for not installing the basic-oxygen converter during the 1950s. However, this criticism lost sight of the fact that severe pressure was brought on the steel industry by government to build capacity before the oxygen converter was considered a proven method of steelmaking.

Steel output reached its peak of 106 million tons in 1955 and remained close to that level for the next two years, when it recorded 105 million and 102 million tons respectively. In the following year, 1958, there was a precipitous drop to 77 million tons, and not until 1964 did output surpass the 1955 high. Thus, for a period of six years, the steel industry in the

United States, unlike that of the rest of the world, operated at two-thirds of its capacity and had no need to install additional facilities.

Oxygen Steelmaking

The oxygen process came into operation in the United States with an initial output of 280,000 tons recorded in 1955. The first installation of the basic-oxygen steelmaking process was made at McLouth Steel Corporation, near Detroit, with a small converter about the same size as the original basic-oxygen converter installed at Linz, Austria. The small-size unit was not acceptable to the U.S. steel industry, and within the next few years larger converters were developed. The second unit to be installed was at the Jones & Laughlin Steel Corporation plant in Aliquippa, Pennsylvania, where an 80-ton converter was put in operation. Kaiser Steel Company at Fontana, California, installed a 100 ton converter in 1958. A quantum leap took place when National Steel Corporation, at its Great Lakes division near Detroit, installed two 300-ton converters in 1962.

By 1963, oxygen steel output had risen to 7.7 million tons. Subsequently, with steel production on the increase and with the need to replace a number of obsolete Bessemer and open-hearth units, oxygen steel production increased substantially year by year. A significant impetus was given to the installation of new capacity by a change in the tax laws in 1962 that established an investment-tax credit that provided added funds to install equipment more rapidly.

Basic-oxygen steel production rose from 7.7 milion tons in 1963 to 70 million tons in 1977 and reached a peak of 76 million tons in 1978. At that time it constituted 62 percent of the nation's steel output, while the open-hearth production had declined to 17.3 million tons, or 14 percent of total output.[14] Further, in adopting the basic-oxygen process, the U.S. steel industry was the first to increase the size of the vessel to produce 300 tons per heat. At the present time, the open hearth is rapidly being phased out and, by 1985, will probably produce less than 10 million tons.

The rapid adoption of the basic-oxygen process in Europe and Japan as compared with the slower pace in the United States can be explained, in part, by the fact that the industry in Japan and Western Europe was growing in the 1960s and the oxygen process was employed in the new plants, whereas the United States had its growth period in the 1950s before the oxygen process was fully developed. The U.S. industry reached the peak of its capacity in 1960 and had no need to expand beyond that level. The installation of oxygen steelmaking equipment, although large after 1960, was almost entirely a matter of replacement.

At the present time, almost every steel mill based on the blast furnace has a basic producing unit employing the oxygen steelmaking process. A number of open hearths are still in operation. However, the rapid replacement of the open hearth by the basic-oxygen converter is evident from the fact that in 1953 the open-hearth process accounted for 90 percent of steel production with no production whatever from the oxygen converter. In 1978, the oxygen converter provided for 62 percent of the output, while the open hearth had declined to the aforementioned 14 percent.[15]

In the years ahead, steelmaking in the United States will be confined almost exclusively to the oxygen converter and the electric furnace. By 1985, these two processes will account for 92 to 95 percent of output, with oxygen steel representing two-thirds of this total and the electric furnace, one-third.

During the past ten years, electric-furnace output has increased from 18.4 million tons in 1970 to a high of 30.8 million in 1979.[16] In 1980, electric-furnace capacity was approximately 36 million tons, due to grow to 41 million tons by 1985. Three major factors are responsible for the installation of this facility, including:

1. Lower capital costs than the basic-oxygen process, which must be supported by blast furnaces and coke ovens;
2. The rapid increase of minimills;
3. The installation by the large integrated producers of electric-furnace units to supplement capacity in smaller amounts than could be achieved by adding more oxygen steelmaking capacity.

Future installations of oxygen steelmaking will be made at those locations where blast furnaces are now supplying hot metal to open hearths and where molten iron is already available without additional investment. Several such open-hearth facilities are in operation throughout the country, so it is conceivable that by 1986 an additional 7 million tons of oxygen capacity will be added to the 1981 total.

Continuous Casting

The U.S. steel industry adopted continuous casting in terms of total tonnage at a pace second only to Japan; yet in terms of the percentage of steel cast, it was considerably slower than a number of other countries. In 1972, 7.0 million metric tons of U.S. raw-steel production were continuously cast. Only Japan, with 16.4 million tons, had a greater amount. The third country was the USSR with 6.9 million tons, and in fourth place was West Germany with 6.1 million tons. Table 5-3 gives the amount of steel continuously cast by Japan, the United States, the USSR, and West Germany from 1972 to 1980. In percentage terms, the United States in 1980 stood at

Table 5-3
Tonnage of Continuously Cast Steel in Selected Countries: 1972-1980
(millions of metric tons)

Year	Japan	United States	U.S.S.R.	West Germany
1972	16.5	7.0	6.9	6.1
1973	24.7	9.3	7.0	8.1
1974	29.4	10.7	7.4	10.3
1975	31.8	9.7	9.7	9.8
1976	37.6	12.2	11.7	12.0
1977	41.8	14.3	12.2	13.3
1978	47.2	17.6	14.4	15.7
1979	58.1	20.9	15.3	17.9
1980	66.3	20.6	18.6 (est.)	20.3

Source: IISI, *Statistical Reports,* No. 125, for years involved (Brussesl).

20.3 percent, against 59.5 percent for Japan, 12.5 percent for the USSR, and 46 percent for West Germany.

The acceptance of continuous casting was slow, due principally to the fact that the steel consumers were not inclined to accept sheets that came from a continuously cast slab. There was considerable difficulty at the outset in producing a uniform quality, and thus a number of the companies hesitated to install the equipment for slab production. This was particularly true of the attempts to cast rimmed steel. The rimming action in the steel, which involved the elimination of gases, was not conducive to the production of a uniform slab in the continuous-casting machine. When the industry abandoned the casting of rimmed steel and substituted a killed steel in its place, the production of continuous casting improved immensely. In the late 1970s it became a more-acceptable type of production than rolled ingot-steel production for some applications. This is particularly true of tin plate.

The U.S. steel industry has a large number of efficient primary mills to produce slabs, and under certain circumstances, the production from these mills is economical and acceptable. Further, the rolling mill serves as a safety valve in the event of difficulties at the casting machine. The presence of such a large rolling capacity has acted, to some extent, as a restraining force on the growth of continuous casting.

At the present time, two U.S. mills of significant size, outside of the minimills, depend completely on continuous casting. They are the United States Steel Corporation plant at Baytown, Texas, and the McLouth plant, near Detroit. Most of the minimills depend completely on continuous casting, as do some of the medium-sized mills.

During the next five years, it is quite possible that the tonnage continuously cast by the U.S. steel industry will be double the 1980 figure of 20.6 million tons. In 1980 and 1981, three continuous-casting units were installed and, thus, were not fully productive for those years. Currently, twelve others are either on order or in the process of construction.

The installation of continuous-casting facilities has top priority on the list of captial-equipment projects in the steel industry. A report by the Office of Technology Assessment, which reviewed technology and steel-industry competitiveness, states "the most important technological change for integrated steelmakers during the next ten years will be greater adoption of continuous casting."[17] This is due to the advantages provided by continuous casting, which include energy savings, higher labor productivity, lower capital costs per ton, and an improvement in the quality of steel.

Direct Reduction

The evolving world development of direct reduction has thus far been characterized by a noticeably incongruous set of circumstances insofar as the United States is concerned. Although technology developed and commercialized by U.S. companies is being used to produce nearly two-thirds of the world's steelmaking grade DRI, much smaller shares of the world's output and installed capacity (approximately 9 and 7 percent respectively, in 1980) are produced by facilities located in the United States.

U.S. companies, including the major steel producers, conducted extensive research into direct reduction before World War II and have developed five of the ten technologies currently being used to produce DRI of steel-making grade. These are (1) Midrex, developed by Midland-Ross and Korf; (2) Armco, developed by Armco Incorporated; (3) SLRN, developed by Republic Steel, National Lead, Stelco, and Lurgi; (4) FIOR, developed by Exxon; and (5) ACCAR, developed by Allis-Chalmers. Taken together, on a world basis, they accounted for 4.6 million tons of output in 1980 out of a world total of 7.2 million tons. Meanwhile, U.S. production continues to be constrained by a widespread lack of economically priced natural gas. In 1980, output totaled 617,000 tons, down from the 1979 level of 798,000 tons.[18]

Largely because of the adverse influence of natural-gas prices, U.S. direct-reduction capacity has remained relatively static since 1973 and probably will not move appreciably above its current level in the near future. To date, four U.S. plants have been built to supply steelmaking-grade DRI, their annual capacities aggregating 1.1 million tons. Three natural-gas-based units, installed prior to the onset of the energy crisis in 1973–1974, account for 95 percent of this capacity, while a fourth unit, a smaller, coal-based demonstration plant started up in 1978, is operated on an intermittent basis to process test batches of alternate feedstocks.

The natural-gas-based plants include:

A Midrex plant, installed by Oregon Steel at Portland, Oregon, in 1969 with a 300,000-ton annual capacity.

A Midrex unit, constructed by Georgetown Steel Corporation at Georgetown, South Carolina, in 1971 with a 400,000-ton annual capacity.

An Armco unit, installed at Armco Steel Works in Houston, Texas, in 1972 with a 330,000-ton annual capacity.

In May 1980, excessive increases in the price of natural gas forced the permanent shutdown of the nation's first commercial plant, the unit installed by Oregon Steel Mills of Portland. For all practical purposes, including the coal-based plant, this reduced U.S. capacity to 790,000 annual tons, even though the Oregon plant remains intact, and no immediate plans have been adopted for its dismantling or disposal. The Armco plant in Texas is in jeopardy since the ten-year gas contract expired in the spring of 1982, and the new price of gas is at least fifteen times as great as the previous price. As of mid–1982, operations had been suspended at both the Armco and Georgetown installations.

In 1980, U.S. steelmakers purchased some 180,000 tons of DRI from producers overseas, and in the years ahead, as their need for DRI increases, largely to improve the charge in the electric furnace, a growing volume of DRI will likely be imported. Anticipating this trend, two major U.S. scrap dealers have formulated plans to market imported DRI to domestic consumers. A supply of imports can be drawn from several potential foreign sources including Canada, Mexico, Venezuela, and Trinidad. A survey of electric-furnace operators making high-quality steels indicates widespread interest in DRI material because of the deterioration in scrap quality.

Although future increases in U.S. direct-reduction activity will have to rely on the use of reducing agents derived from coal, the coal-based processes currently in operation on a worldwide basis usually number among the least commercially and technically advanced, even though some have long histories of development work and operating experience. In 1980, the world's eleven coal-based facilities produced an aggregate of 372,000 tons of steelmaking-grade DRI, representing just 30 percent of their combined rated capacity of 1.27 million tons. Given their predominant reliance on rotary-kiln techniques, the coal-based processes have thus far generally shared common technical and operating limitations that have discouraged their wider acceptance. Future progress in the adoption of direct reduction in the United States will require further development and commercialization of more-effective coal-based technologies. With respect to coal-based direct reduction, the Office of Technology Assessment in its steel-industry report placed high hopes on this process stating "the technological development with the greatest advantages and best possibility of limited commercial adoption within 5 to 15 years is coal-based DR of iron."[19]

Other Improvements

As is the case in a number of other steel-producing countries, the U.S. industry has developed and applied numerous other technolgical advances in the past twenty years. One of these, which might be considered a breakthrough, is the beneficiation of low-grade ore whereby 20–30 percent iron-bearing material is upgraded and formed into pellets with 60–65 percent iron content. This upgraded burden has improved blast-furnace practice and increased production considerably so that more iron is made with fewer furnaces.

In electric-furnace steelmaking, the use of ultrahigh power has reduced the time required per heat of steel and thus increased the output of individual furnaces. Transformers of 100,000 KVA are used on the larger furnaces. Further, the installation of water-cooled side panels and, recently, water-cooled roofs has increased the working volume of the electric furnace by 10–30 percent, depending on the type of unit, and has improved the lining life.

In the cokemaking process, a consortium has developed a plant for formed coke that produces some 225 tons per day. This has been used in limited tests, but the construction of a production-size plant capable of feeding large blast-furnaces is a matter for the future.

The application of computer technology to various steps in the steel-making and rolling processes has also improved product quality and increased productivity in the United States as it has in other countries throughout the world. Process control by computer was the subject of a paper given at the International Iron and Steel Institute meeting in Toronto in October 1981. It covered the subject fully and indicated the scope of computer uses in the steel industry. In referring to the breadth of computer application, the paper stated:

> At the present time, there is no area of a steel plant where someone somewhere does not have a computer for some aspect of control . . . they range from coal mines, ore beneficiation, through coke, iron, and steelmaking to rolling, finishing, warehousing, and shipping.[20]

In the next few years, with no expansion plans for steel-melting capacity, large sums will be invested in improving technology. This will mean a wider adoption of the aforementioned improvements, as well as the installation of continuous-annealing lines for sheets.

Ownership

All of the steel companies, large and small, integrated and nonintegrated, in the United States are privately owned. However, a change has occurred in the private ownership of many companies in the last fifteen years. In the late 1960s, a number of steel companies were taken over by conglomerate organizations. These included five integrated operations, all of which were absorbed in 1968: Jones & Laughlin was acquired by Ling-Tempco-Vought; Youngstown Sheet and Tube was acquired by Lykes Steamship Company; Crucible Steel became a part of Colt Industries; CF&I was acquired by Crane; and Sharon Steel was acquired by NVF. Lone Star Steel, which was first acquired by Philadelphia Reading, became a part of Northwest Industries in 1965.

Other takeovers included Jessop Steel by Athlone, Firth Sterling and Columbia Steel by Teledyne, and Continental Steel by Penn-Dixie Cement.

In 1977, Youngtown Sheet and Tube closed approximately half of its facilities and faced bankruptcy. In 1978, it was acquired by Jones & Laughlin to save the remaining facilities that were located at Indiana Harbor, near Chicago. This was permitted by the Justice Department under the Failing Company Doctrine. An attempt was made by a local group in Youngstown, Ohio, to acquire the facilities that Youngstown Sheet and Tube had closed down in that area, but this was not successful. It was to have been a community-owned steel company, paid for with federal funds, and the closest thing to a government-funded operation the United States would have had.

Trade

Paradoxically, the United States, which was the world's leading producer of steel during most of the post–World War II period, even as it held that position, became the world's largest importer of steel. Imports rose gradually in the first half of the 1960s, after which time they increased rapidly to a peak for the decade of 16.3 million tons in 1968.

From the turn of the century through 1958, the United States had been a net exporter of steel; however, in 1959, when a four-month steel strike crippled production, the United States became a net importer with 4.1 million tons of imports as opposed to 1.4 million tons of exports. This was regarded as a one-time unusual situation attributable to the strike and, thus, generated little concern as a long-term possibility. Such reasoning was quite logical in view of the fact that two years earlier, in 1957, the United States had exported 4.7 million tons as opposed to 1 million tons of imports, almost a five-to-one favorable ratio.

For the decade prior to 1957, U.S. steel products were in demand in the industrialized countries throughout the world as they rebuilt from the ravages of war. However, despite the worldwide demand for U.S. steel during this period, a relatively small amount was exported since domestic demand in most years was strong enough to absorb virtually all of the steel that could be produced. Further, although a net exporter for years prior to 1959, the U.S. steel industry, with the exception of United States Steel Corporation, was never particularly export minded since the home market, except for a few depression years, was usually large enough to consume the industry's output.

During the years from 1960 to 1980, exports constituted a very small portion of the U.S. steel industry's shipments. With the exception of a few years, they ranged between 1.5 million and 3.6 million tons. In one year, 1970, they reached 6.5 million tons; however, this still constituted less than 8 percent of steel-industry shipments. Table 5–4 indicates the great disparity between exports and imports during the past twenty years and also gives the percentage of total steel-mill shipments that exports constituted.

Since the strike year of 1959, the United States has been a net importer of steel. The reasons for this surge in import tonnage during the 1960s were:

The rapid and continuing increase of steelmaking capacity in Western Europe and Japan, providing increased steel for export;

Low costs of production, due to new mills and low wage rates in Western Europe and Japan, that allowed these producers to offer steel in the United States at prices substantially below the U.S. level;

The increased price-competitive position of overseas steel producers resulting from the relative over-valuation of the U.S. dollar, which was officially linked to gold and retained its exchange value despite a deteriorating U.S. balance-of-payments position, particularly during the second half of the decade;

The fact that the triennial labor negotiations between the steelworkers union and the steel industry always presented the possibility of a strike and that steel consumers, mindful of the 1959 experience, bought steel from abroad to hedge against this possibility in the negotiating years.

During the labor negotiations in 1965, steel buyers looked to foreign suppliers to protect themselves against the possibility of a strike, and imports rose from 5.8 million tons in 1964 to 9.4 million in 1965, the largest

Table 5-4
Imports and Exports of Steel Products for the United States: 1960–1980
(millions of metric tons)

Year	Imports	Exports	Exports as a Percentage of Total Shipments
1960	3.1	2.7	4.2
1961	2.9	1.8	3.0
1962	3.7	1.8	2.8
1963	4.9	2.0	2.9
1964	5.8	3.1	4.0
1965	9.4	2.3	2.6
1966	9.8	1.5	1.8
1967	10.4	1.5	2.0
1968	16.3	2.0	2.3
1969	13.0	4.7	5.5
1970	12.2	6.4	7.8
1971	16.6	2.5	3.2
1972	16.1	2.6	3.2
1973	13.9	3.7	3.7
1974	14.5	5.3	5.3
1975	11.8	2.7	3.8
1976	13.0	2.4	3.0
1977	17.4	1.8	2.2
1978	19.1	2.2	2.5
1979	15.9	2.5	2.8
1980	14.1	3.7	4.9

Source: American Iron and Steel Institute, *Annual Statistical Reports* (Washington, D.C.: various years).

increase in any one year up to that time. In the two subsequent nonnegotiating years, the level was slightly increased because, as steel customers found the quality of imports satisfactory and the prices attractive, some of them continued to commit a certain portion of their requirements to imported steel. Total imported tonnage in 1966 was 9.8 million tons, increasing in the following year to 10.4 million.

In 1968, steel consumers were faced with another negotiating year, and discussions between labor and management aroused fears of a strike. Accordingly, the steel buyers bought heavily from foreign suppliers, with the result that imports reached an all-time high of 16.3 million tons, an increase of 5.9 million tons in one year. As a result of these increases during the 1960s, Western European and Japanese producers came to look on the U.S. market as a regular, dependable outlet for their products and began to talk about their share of it.

The surge in imports in the mid to late 1960s started a movement among the U.S. steel producers and the steelworkers union to limit the import tonnage. The first device that was adopted involved voluntary restraints on the part of exporters to the U.S. market. After these were discarded in 1973, there were no restraints until the trigger-price mechanism was imposed in early 1978. Antidumping suits were filed against foreign producers in early 1980, resulting in the suspension of the trigger price. It was reinstated later that year but suspended again in 1982 as more antidumping suits were filed. A full discussion of these events is given in the chapter on international trade.

Impact of Steel Imports on the United States

The increase in steel imports during the past twenty years has had a multiple impact on the United States. Among other things it has:

1. Increased steel-market competition by providing consumers with access to alternative sources of supply and with cheaper steel, except during periods of peak demand.
2. Increased the pressure to replace and modernize steel-industry facilities and to improve the quality of steel products.
3. Influenced the tax policies of the U.S. government to make additional funds available for plant and equipment modernization.
4. Changed the basic structure of labor-management relations in steel and resulted in the conclusion of a no-strike labor agreement.
5. Restricted the growth of the U.S. steel industry by displacing domestic production, forcing the cancellation of plans to construct additional capacity, and contributing to the shutdown of existing capacity.
6. Resulted in the loss of employment both in the steel industry and in a variety of supplier industries.
7. Depressed steel-industry earnings by reducing output and operating rates, increasing fixed costs per ton of production, and periodically disrupting the U.S. price structure.
8. Increased the incentive for steel companies to diversify into nonsteel activities.

As import tonnage began to grow in the mid-1960s, reaching 9.4 million tons in 1965, plans for the expansion of facilities in the United States were seriously curtailed. Mention has been made of the canceled plans for integrated steel mills on the West Coast by United States Steel, Bethlehem Steel, National Steel, and Armco. Had their plans materialized, they would have probably added an aggregate of 7.5 million to 9 million tons of steelmaking capacity on the West Coast. Import penetration, particularly from Japan, was a dominant motive in these cancellations because, as early as 1965, the Japanese looked on the West Coast as part of their natural market. The Sparrows Point plant of Bethlehem Steel Corporation, located near Baltimore, Maryland, has shrunk in size due to steel imports on the West Coast, since it supplied sizeable tonnages to that market by shipments through the Panama Canal.

Steel capacity in other areas was also adversely affected by import penetration. This was particularly true of the Great Lakes region, which received about one-third of total U.S. imports in years when they were high, such as 1965, 1968, 1971, 1977, and 1978. There is little doubt that these large tonnages of imports were a factor in the closure of plants, such as the Youngstown Sheet and Tube plant near Youngstown, Ohio; part of the Bethlehem Steel plant at Lackawanna near Buffalo, New York; and the United States Steel Corporation plant in Youngstown, Ohio. Most of the facilities in these plants were obsolete, because the lack of demand for the products they produced led the companies to decide against replacing this equipment. Plans for new integrated, greenfield-site plants in the Great Lakes area were canceled due to the high cost of installation and the financial position of the companies involved. Steel prices would not permit the return needed on the huge investment required to build greenfield-site plants.

Hope for expansion of the U.S. steel industry was dimmed partly by the steady flow of large import tonnages, which means that steel producers in the 1980s cannot look to the entire U.S. market for the sale of their product but must look to 85 percent of it, since some 15 percent seems to have been permanently lost to imports.

Another impact of imports on the U.S. industry was that they increased the pressure to institute modernization and replacement programs. They were undertaken in the early 1960s and significantly aided by the passage of the Revenue Act of 1962 that provided a 7 percent investment-tax credit making additional funds available to modernize plants and equipment. The purpose of the investment-tax credit was to combat imports by modernizing the U.S. industry to make it competitive with the Europeans and Japanese, both of whom had rebuilt and expanded their industries with new equipment after World War II. In recommending the investment-tax credit, President Kennedy stated:

> Today, as we face serious pressure on our balance of payments position, we must give special attention to the modernization of our plant and equipment. Forced to reconstruct after wartime devastation, our friends abroad now possess a modern industrial system helping to make them formidable

competitors in the world markets. If our own goods are to compete with foreign goods in price and quality, both at home and abroad, we shall need the most efficient plant and equipment.[21]

Thus, imports had a decided effect on U.S. tax policy that provided tax credits for the steel industry to install modern equipment. In the 1960s, after the passage of the Revenue Act, the installation of the basic-oxygen process was accomplished rapidly. In 1963, production by this process was 7.7 million tons, and by 1979 it had reached a high of 76 million tons, because of the installation between 1962 and 1979 of 68 million tons of basic-oxygen steelmaking capacity. During the 1960s and early 1970s, twelve hot-strip mills were built. These were so much improved over their earlier counterparts that they were referred to as the second generation of hot-strip mills. Examples of modernization could be multiplied.

It should be noted that imports have been closely tied to politics since the last half of the 1970s. When antidumping suits, in keeping with the U.S. trade laws, have been threatened or filed, politicians from those foreign countries involved have exerted pressure on the Washington administration at the highest level to stop the suits and gain favorable conditions for their imports into this country. Some of these countries have objected to any sort of protection that restricted their exports, while indulging in significant protective measures in their own markets.

Steel imports have had a major impact on the U.S. steel industry's labor relations. In those years when the industry and the union negotiated a new contract, steel consumers, mindful of their four-month strike in 1959, resorted to building up their inventory in fear of a possible strike. In the negotiating year, 1965, import tonnage increased by 65 percent over the previous year and, in 1968, another negotiating year, imports increased by 50 percent. Recognizing that this was depriving the U.S. steelworkers of jobs, the union and management concluded the Experimental Negotiating Agreement in 1973, which was a landmark in steel labor relations, as it provided arbitration for unresolved issues and rendered an industrywide strike impossible. In reference to the new agreement, the annual report of the United States Steel Corporation for 1973 stated:

Because of the fear of a work stoppage at labor contract time every three years, it has long been the practice of customers to build massive steel inventories prior to the Steelworker's contract deadline and to deplete them either during the subsequent strike period or after a new agreement has been signed. In addition, even though there has been no industry-wide strike since 1959, some customers have increasingly contracted with foreign suppliers for their steel needs to assure a continuity of supply and to supplement the building of inventories. Thus, penetration of imported steel into American markets has surged in each contract negotiation year since 1956. Likewise, steel operations and employment have risen rapidly to very high levels before the contract deadline and then have dropped to very low levels afterwards—both very inefficient extremes. These conditions have been harmful to steel companies, employees, stockholders, steel communities and to the nation. We believe that this agreement, arrived at voluntarily by the parties, will be of great benefit in alleviating the undesirable conditions previously experienced.[22]

The Future of the U.S. Steel Industry

Due to the current overcapacity in the world, there is no prospect for growth in the U.S. steel industry by 1985. The industry will actually shrink by as much as 8 million to 10 million tons of raw steelmaking capacity during the next few years, since there are no plans by the integrated companies for expansion and some will close down plants or parts of plants. The Kaiser Steel Corporation has already written off its coke ovens, blast furnaces, and 3 million tons of steelmaking capacity. McLouth Steel, with approximately 2 million tons of capacity, is in a precarious position, and it is certain that several obsolete facilities, such as the Homestead open-hearth shop of United States Steel, will be closed down, not to be replaced. Other mills whose capacity could well be reduced include the Lackawanna plant of Bethlehem Steel as well as the Weirton plant of National Steel, which is in the process of being sold to the employees.

One or two minimills, such as the 500,000-ton plant announced by Timken, will be built and, to a limited extent, this will neutralize other losses. By 1985, the industry in the United States will have a reliable steel-making capacity between 127 million and 129 million tons. This shrinkage will also mean fewer jobs and a decline in total steel employment. Between 1974 and April 1981, total employment, including salaried and hourly paid workers, dropped from 512,000 to 403,000. The latter number is significant since the operating rate in that month was over 85 percent of capacity. Thus, the loss of jobs was permanent. Since early 1981, employment has fallen to approximately 310,000 due principally to temporary layoffs because of the drastic drop in the operating rate. Many of the laid-off employees will be rehired. However, because of the projected shrinkage in capacity, the industry through 1985 does not represent a growth potential for employment.

Location

There will be no significant growth in any of the steel-producing areas during the remainder of the decade, nor will there be a dramatic shift in steelmaking capacity as some had thought. The Pittsburgh area will decline as some obsolete plants are closed, not to be replaced. Chicago, which is the dominant steel producer, will maintain its current level of capacity, while plant closures will affect Detroit, the Southeast, and the West Coast.

The expansion in the southwest, which took place in the past 10 years, has reached its limit and there will be no further significant increase for at least five years. Most of this capacity, with the exception of Lone Star's integrated plant, is dependent on the electric furnace and, while there may be some increase in furnace size in the smaller plants, no major installations are expected.

The west coast, which has always been a steel-deficit area, will be more so when Kaiser closes down 3 million tons of steelmaking capacity in 1983. In the southeast, the United States Steel plant near Birmingham, Alabama, may well be scaled down; however, no final decision has been made on this matter. The plants on the east coast will remain at their 1982 capacity with the exception of the Fairless Works of United States Steel, which will be somewhat upgraded.

There is no indication that the steel industry will construct additional large facilities on the seacoasts. At present there are two major plants at tidewater locations: the Sparrows Point plant of Bethlehem Steel in Baltimore and the Fairless Works of United States Steel Corporation near Philadelphia. The large plants on the Great Lakes have some of the advantages of a deepwater location such as the ability to receive iron ore by low-cost water transportation.

Comparisons have been made between the Japanese and U.S. steel industries with respect to plant locations. The Japanese plants are located on the seacoast and can readily receive imports of higher-grade iron ore and coal and are also in a position to export steel products. This seacoast location is necessary in Japan due to the fact that virtually all of its raw materials must be brought in from abroad and huge quantities of steel are shipped overseas. Seacoast locations in Europe are also becoming more important, as a greater portion of the steel industry's raw materials is imported and more steel is exported.

In the United States the need for a seacoast location is by no means urgent, for most of the raw materials are available domestically and a relatively small percentage of the steel produced is shipped abroad. Nearly all of the industry's coal comes from the interior of the United States and is shipped by rail to the various steelmaking units. Two-thirds of the ore consumed by the U.S. steel industry is produced domestically, much of it in the Great Lakes area, whence it is shipped to the industry's plants located on the Great Lakes, as well as those at other inland locations. Of the ore imported, over 50 percent comes from Canada via the Great Lakes. Further, since 95 percent of the steel-mill products of the U.S. steel industry are consumed at home, many of the plants located inland are close to the consuming industries. This is not likely to change in the next five or more years, so there is very little to be gained by a shift to coastal locations.

Some argue that a coastal location would provide access to richer ores than are available in the United States. This is true, but the savings from the use of such ores would be minimal compared with the cost of relocating steel plants.

The Chicago area is by far the dominant steel producer in the United States. Its tonnage is far greater than that of any other area whether the industry operates at a high or low rate of capacity. The following table gives the breakdown of steel production on a weekly basis for all of the areas of the country at an 89 percent rate of operation and at a 59 percent rate of operation.[23]

	Week of 3/28/81 89.2 Percent Rate of Operation (thousands of net tons)	Week of 3/12/82 59 Percent Rate of Operation (thousands of net tons)
Northeast coast	314	197
Buffalo	73	32
Pittsburgh	464	250
Youngstown	154	92
Cleveland	182	113
Detroit	193	95
Chicago	668	513
Cincinnati	106	87
St. Louis	101	77
Southern	265	169
Western	151	113

In comparing Chicago with Pittsburgh at the two rates of operation, the drop was far more pronounced in Pittsburgh than in Chicago. In Pittsburgh, the drop from 464,000 to 250,000 net tons was 46 percent, while the Chicago drop from 668,000 to 513,000 represented a 23-percent decline. Six of the seven large, integrated companies, including United States Steel, Bethlehem, Jones & Laughlin, Inland, Republic, and National have plants in the Chicago area. There are no current plans to increase this concentration by shifting steel capacity to Chicago. Some plants in other areas will be reduced in size and this will increase the percentage of concentration in Chicago, but not the actual tonnage.

Production Processes

During the next five years the open-hearth process will continue to decline and output will fall to about 10.0 million tons. The remaining production will come from the basic-oxygen converter and the electric furnace. Oxygen-converter capacity in the United States at the present time is some 77.0 million to 79.0 million tons, with approximately 36.0 million tons of electric-furnace capacity. The electric-furnace capacity will increase as open hearths drop out and, by 1985, will be approximately 41.0 million tons. Basic-oxygen capacity will increase, where it can replace open-hearth facilities that are supported by blast furnaces.

The electric-furnace process will improve as technological advances, such as water-cooled panels and roofs, ladle refining, and the use of oxygen and ultrahigh power, are applied. This will help to increase the furnaces' capacity without necessitating adding more units. The limiting factors on the growth of the electric furnace are the cost of electric power, which has increased fivefold in some areas during the past ten years, and the availability of quality scrap.

Minimills

The electric furnace has been the means by which the minimill came into operation in the U.S. steel industry, and many feel that this type of facility will replace a significant portion of the integrated steel capacity. An analysis of the minimill concept and projections for its future will shed some light on this question.

The minimill, developed during the 1960s, consisted of a small electric furnace, very often a continuous caster, and a bar mill that produced principally concrete reinforcing bars and other commercial-quality bars. The annual capacity of these mills during the 1960s ranged from 40,000 tons up to 300,000 tons. Since that time many of these mills have expanded, so that a number produce more than 500,000 tons. In the cases of Florida Steel and Nucor Corporation, which operate several minimills, total capacity for each company approaches 2.0 million tons.

The term, minimill, seems to have outlived its usefulness, for a number of companies never considered minimills, for example, Atlantic Steel at Atlanta, Georgia, and Continental at Kokomo, Indiana, are now in the same tonnage class with mills that were originally considered minimills. The Korf mills at Georgetown, South Carolina, and Beaumont, Texas, are each capable of producing more than 600,000 tons of raw steel. The new Bayou steel mill in Louisiana has approximately the same capacity, as does the Raritan River Steel Company mill recently erected in New Jersey. Some of these mills are close to the capacity of Laclede Steel, an electric-furnace operation near St. Louis, Missouri, which was never considered a minimill. Thus, it seems more appropriate to speak of semi-integrated, electric-furnace plants rather than minimills. Taken together, this grouping would include such operations as Northwestern Steel and Wire, Lukens Steel, and Phoenix Steel, as well as all the smaller independents. The total capacity of these plants is between 19 million and 20 million tons, or approximately half of the electric-furnace capacity in the country. The other half is operated by the integrated mills.

The possibility of many semi-integrated, electric-furnace companies, which with the exception of some larger producers have generally confined their efforts to the production of bars and small structural sections, moving into the production of sheets is at present quite remote. This is so in spite of the fact that facilities have been developed that could be installed by the small electric-furnace operators, to produce limited amounts of narrow strip, about twenty-four inches wide. The production of wide sheets cannot be accomplished without a modern strip mill, requiring an investment of at least $300 million. Further, the integrated mills have tremendous capacity for producing sheets and it is unthinkable that the semi-integrated, electric-furnace operators would move into this area. In addition, sheet products very often must have a deep-drawing quality and it is extremely difficult to make such products from scrap that has an unacceptable percentage of tramp elements or residuals. Semi-integrated, electric-furnace operations, with one or two exceptions, will not produce sheets in any quantity for the foreseeable future, but they have a definite position in the U.S. industry.

Capital Expenditures in the 1980s

In the spring of 1981, the U.S. steel industry, in a state of euphoria due to high operating rates and substantial profits, planned to spend a total of $5.5 billion on capital projects. This did not include any expansion of steel-melting facilities, but concentrated on the replacement of obsolete plants with modern, competitive equipment. As the year wore on, the euphoria turned to grim pessimism, with operating rates dropping from 89 percent of capacity in April to less than 60 percent in the late fall and early winter.

Early 1982 brought no relief as steel demand declined further during the first quarter, resulting in substantial losses for most of the steel producers. As a result, most of the major scheduled programs were either postponed or canceled. One of these was the $670 million seamless tube mill and continuous-casting unit planned by Armco. The project has been postponed indefinitely due to the drop not only in general economic activity, but in oil-drilling rigs operating. Other programs involving the relining and modernization of blast furnaces have been put on hold. Another sign of the cutback in capital expenditures was the announcement by National Steel Corporation that it wished to divest itself of the Weirton plant with its 3 million tons of steelmaking capacity. The reasons given were that the company did not wish to invest the large sums needed to put this plant in a competitive position.

Projects that have been continued include part of the $750 million program of Bethlehem Steel, as well as the $600 million program of United States Steel Corporation involving the installation of a continuous caster and seamless pipe mill in Alabama.

During the remainder of 1982, losses are expected to mount, and this will affect the capital programs of many steel companies and will lead to a more judicious selection of facilities to be installed. There will be a moratorium on additional capacity, and this is highly desirable for the economic viability of the industry, since the cut in capacity and even a modest increase in demand will permit higher operating rates.

The technology to improve productivity and profitability is available. The basic problem is the availability of capital to install that technology. At present, it is difficult to see how the industry can maintain itself in a competitive position without a minimum annual expenditure of $2.5 billion on steel facilities. The funds will be drawn from depreciation and, hopefully, retained earnings. If the earnings are not forthcoming, it will be necessary to borrow the funds, however, not every company will be able to borrow the sums needed.

Problems Facing the U.S. Steel Industry

Two factors cloud the future of the steel industry in the United States during the 1980s: one is the future of the automotive industry and the second is the degree of steel-import penetration. On the bright side, there is the possibility of developing new outlets for steel, particularly in the field of energy and from the need to replace the country's infrastructure.

The automotive industry has constituted as much as 22 percent to 25 percent of the steel industry's market in a number of years during the post-World War II period. This was particularly true in the mid 1950s and in the mid 1960s, as well as the late 1970s. In some years, automotive-industry consumption of steel was in excess of 22.0 million tons, when shipments of steel products to the automotive industry from other industries, such as fasteners and Service Centers, are included.

In 1980, when automotive-industry production fell by 30 percent compared with the 1979 high, dropping from 11.5 million vehicles to 8.0 million, and with the production of lighter cars, it consumed about 11.0 million tons, or approximately 15 percent of the steel industry's output. The automotive industry's consumption of steel in 1981 was not much higher than in 1980, and prospects for 1982 are less than sanguine. Table 5-5 gives the number of vehicles produced by the automotive industry, steel ship-

Table 5-5

U.S. Automobile Production, Total Steel Shipments, and Sheet and Strip Shipments to the Automotive Industry: 1955-1980

Year	Millions of Vehicles Produced	Steel Used (million tons)	Percentage of Total Steel Shipments	Sheets and Strip		
				Total Shipments (million tons)	Used in Autos (million tons)	Percent Used in Autos
1955	9.2	17.0	22.1	29.4	12.2	41.7
1960	7.9	13.2	20.5	26.3	10.1	38.2
1961	6.7	11.4	19.1	20.7	8.5	41.1
1962	8.2	13.8	21.5	25.9	10.3	39.6
1963	9.1	15.3	22.4	28.1	11.4	40.6
1964	9.3	16.7	21.6	31.0	11.9	40.1
1965	11.1	18.2	21.7	33.3	13.6	41.0
1966	10.4	16.3	20.0	32.2	12.1	37.6
1967	9.0	15.0	19.7	29.6	11.2	38.1
1968	10.8	17.5	21.0	33.2	13.2	39.9
1969	10.2	16.6	19.5	34.6	12.6	36.4
1970	8.3	13.2	15.9	31.8	10.0	31.3
1971	10.7	15.9	20.1	32.3	12.2	38.1
1972	11.3	16.5	19.8	36.2	12.6	35.0
1973	12.7	21.0	20.8	44.8	16.5	36.8
1974	10.1	17.1	17.3	40.8	13.2	32.5
1975	9.0	13.8	19.0	27.9	11.0	39.3
1976	11.5	19.4	23.9	38.3	15.6	40.6
1977	12.7	19.5	23.6	37.8	15.7	41.6
1978	12.9	19.3	21.7	39.6	15.4	40.0
1979	11.5	16.9	18.5	39.5	13.7	34.6
1980	8.0	11.0	14.5	30.5	9.1	29.6
1981	7.9	11.9	15.0	33.6	9.9	29.5

Source: Motor Vehicle Manufacturers Association, *Motor Vehicle Facts and Figures* (Detroit: 1980), p. 10.

ments received by that industry, and the percentages they represent of all steel shipments. It also gives the total production of steel sheets and strip as well as the percentage consumed by the automotive industry.

During the last ten years, automotive imports have been in excess of 2.0 million vehicles each year. In 1974 imports rose to a high of 2.6 million, declined to 2.1 million in 1975, and then resumed an upward trend, surpassing 3.0 million vehicles in 1978 and 1979. In 1980, the figure was 3.6

million and with a declining market in the United States this had a tremendous impact on domestic automotive sales. As can be seen in table 5–5,
automotive production in the United States fell to 8.0 million units in that
year. Consequently, the automotive industry has been severely affected and
this has had an impact on the steel and other industries that supply automobile producers. In 1981, the production figures of the U.S. automobile
industry were the lowest in twenty years, as they dropped to 7.9 million
while imports declined only slightly. In 1980 and 1981, imports represented
23 percent and 21 percent, respectively, of the U.S. market. The major portion of the imports, about 70 percent, came from Japan, and until the Japanese automakers restricted shipments to the United States in 1981 by some
180,000 units, they showed little disposition to curtail sales to the American
market.

There is much concern that the large volume of Japanese imports will
become a permanent fixture and will have a lasting adverse effect on the
automobile industry and its suppliers. This concern is evident throughout
the steel industry, where huge investments were made in equipment during
the 1960s and 1970s to supply the automotive industry. In the thirteen-year
period from 1961 to 1974, twelve new large fully automated hot-strip mills
were installed. These produced sheets and, as table 5–5 indicates, the automotive industry is the principal consumer of sheets. In 1980, the amount of
sheets consumed by the automotive industry dropped to less than 30 percent
of steel-industry sheet output. When one considers that traditionally it was
much closer to 40 percent, the concern on the part of the steel industry is
understandable.

The twelve new hot-strip mills installed by the steel industry have an
aggregate capacity of 41.0 million tons. In addition to these new mills,
many of the twenty-five older units were rebuilt and modernized, so that by
1974 hot-strip mill capacity in the United States was in excess of 82.0 million
tons. Since that time, several of the older mills have been decommissioned,
so capacity is now in the area of 64.0 million to 68.0 million tons. This far
exceeds demand and, if the automotive industry fails to revive, the implications for steel are less than optimistic. The automobile industry will have to
recover to a production of 11.0 million vehicles for the strip mills of the
steel industry to operate at a high rate of capacity. It is interesting to note
that the decline in automotive production in the United States has affected
not only the domestic steel industry but also the overseas producers who
have had a significant share of the U.S. automotive steel business.

A number of analysts project a recovery in the U.S. automotive industry by 1985 to a minimum of 11.5 million vehicles, and some put the figure
as high as 12.0 million. If this is achieved, steel shipments to the automotive
industry will increase over the 1980 and 1981 levels. However, because of
the smaller cars, the total will not reach the levels achieved in 1973, 1977,
and 1978.

The downsizing of automobiles has had a significant effect on steel-industry shipments. In 1960, the automotive industry used, on an average, 1.77 tons of steel in the production of a car. By 1976, this had dropped somewhat to 1.6 tons per passenger car. In 1981, the figure had declined by 32 percent to 1.1 tons and in 1984 it might be as low as 1.0 ton per car. However, most producers feel that 1.1 tons will be sustained into the indefinite future. Trucks require the purchase of 2.3 tons of steel per unit, down from 3.2 tons in the mid 1970s.[24]

The possibility of growth in the automotive industry in the United States is severely limited. As of 1979 there were 1.5 persons per vehicle and 1.9 persons per passenger car. This is a high degree of saturation and indicates that growth will be minimal. As of 1980 there were 159 million vehicles registered, of which 123 million were passenger cars. Although the growth possibilities are limited, the replacement market is enough to keep the U.S. steel industry operating at a very high rate of capacity. If one assumed an 8 percent scrappage per year, this would mean that 9.8 million passenger cars would be replaced each year. The question arises as to what portion of that total will be imports. If imports are reduced to 15 percent, this would allow 8.5 million passenger cars to be produced by the U.S. automakers.

The second serious problem that the steel industry faces is the degree of import penetration from foreign steel producers. For the years 1976 through 1980, imports averaged 15.4 million tons per year, which was slightly more than 16 percent of the American market. In the latter part of 1981, as imports increased and the domestic market slumped badly, market penetration rose in some months as high as 24 percent and 25 percent. For 1981, imports of 18.1 million tons constituted 19.2 percent of the U.S. market.

The heavy volume of imports during the last quarter of 1981 and the first quarter of 1982 was a significant factor in the decline of the U.S. steel industry's operating rate to less than 50 percent of capacity. It is important to note, however, that they were not the only factor. Other factors included increased operating costs, the obsolescence of some steel-mill equipment, and the general poor state of the economy.

The 1981-1982 experience of heavy import penetration during economic recession tends to obscure the fact that in prosperous years, the U.S. economy needs imported steel, since the domestic industry's capability is inadequate to meet the market's full demand. In fact, the United States is the only major steel-producing country in the industrialized world that does not have the capability to fill its total domestic needs. The present capacity of the U.S. industry in terms of shipments is approximately 96-million metric tons, and, in a good year, the economy could consume between 105 million and 107 million tons of steel products. This, however, is predicated on the automobile industry producing 11 million vehicles. It has fallen considerably below that level and if it does not recover total steel requirements in the economy will be somewhat reduced.

It is safe to say that the United States needs at least 10 million tons of steel imports when its economy is working at a fairly good level. This was not the case in 1980, when steel-industry shipments declined to 75-million tons and imports were 14.1 million tons. With the exception of oil-country goods including casing, drill pipe, and tubing, which were in strong demand throughout 1980 and could not be completely supplied by U.S. producers, few imports were actually needed. In 1979, however, when the domestic industry shipped 91.3 million tons of products to a strong economy, there was a need for at least 10-million tons of imports. In that year, the automotive industry produced 11.5 million vehicles and took some 18 million tons of domestic steel.

During the past few years, imports averaged 15.4 million tons annually, which was approximately 5 million tons in excess of the economy's needs. This excess tonnage, which in some years rose to 9 million tons, depressed the price of imports and also had an effect on the domestic price structure. In 1977, when shipments were 82.9 million tons, which was considerably below the industry's supply capability, the import figure stood at 17.4 million tons. With 1.8 million tons of exports, there was a total availability of steel in the United States of 98.5 million tons, which was more than the economy's requirements. To dispose of this tonnage, prices were slashed to the point where imports were offered at as much as $100 per ton below the U.S. list price. Partly as a result of this, profits for the industry disappeared, and some capacity was permanently shut down.

As a result of this highly competitive situation, steel consumers enjoyed lower prices. This led some observers to maintain that the United States should import as much cheap, subsidized steel as possible to benefit the consumer and permit producers of steel-containing products to better compete in the world market. This may be an attractive short-run strategy, but in terms of a continuing trade policy, it represents very poor reasoning. If low-priced, subsidized steel comes into the United States in large tonnages, thereby affecting the price structure and consequently the profit position of the domestic steel producers, it will force them to close capacity.

Between 1977 and 1981, the U.S. steel industry shut down 12.5 million tons of integrated steel capacity. If these closures continue, the capacity of the U.S. industry will be far below what is necessary to fill the needs of the economy and many more tons of imports will be required. Under these circumstances, when imports are needed and do not enter the market in competition with domestic steel but as a necessary supplement to inadequate domestic supplies, there will be no bargain-basement prices. Importers, recognizing that their products are in demand, will charge what the traffic will bear as they did in 1973 and 1974. An example will illustrate this point. In June 1972, steel plates were sold in the United States for $119 per ton f.o.b. a Western European mill. When the boom reached its height in mid-1974, these same plates were sold in the U.S. market at $440 per ton f.o.b. the Western European mill.

There is little doubt that this will happen again, since the EEC steel producers have suffered losses of hundreds of millions of dollars between 1975 and 1982 and, if the U.S. market improves so that more imported steel is needed, these producers will try to recoup some of their losses by charging higher prices. Thus, to advocate that the United States should accept all of the low-cost, subsidized, foreign steel that it can import, and drive some of our domestic mills out of business, will in the long run (and this could be three to four years) result in much higher prices for steel to the American consumer.

It has been argued theoretically that these higher prices will increase profitability and lead to the installation of more steel capacity by companies in the United States and thus the situation will be rectified. This is doubtful for two reasons: (1) the installation of steel capacity has become very costly and, even with higher prices, the return on new capacity will not be high enough to justify its installation, and (2) steelmakers, remembering the boom of 1973 and 1974, which ended in a drastic drop in 1975, will look for at least three consecutive years of highly profitable operations and a prospect of continued steady demand before making a decision to increase capacity. After the decision, a minimum of two to three years will be required to construct the new facilities and put them in operation and, thus, there will be a prolonged period before new capacity is available.

In addition to their long-run detriment, below-cost imports hold no guarantee of net short-run benefit. While it is true that steel users are afforded the temporary advantage of lower steel prices, any benefit that may ultimately accrue to their customers, employees, or the economy in general must be weighed against the negative short-term impact of losses in steel-industry production, employment, sales revenues, wages, and tax revenues. Additional short-term penalties result from comparable losses in a long list of steel-supporting industries, namely, those engaged in such activities as iron ore and coal mining, steel-plant engineering and construction, and in the production of such consumables as ingot molds, refractories, electrodes, and steel-mill lubricants. The penalties are most often initially perceived as economic, but in their most serious form they affect the social well-being and living standards of individuals, families, and entire communities.

It is imperative that some solution to the steel-import problem be found. Such a solution must be comprehensive enough to permit needed steel imports that stimulate healthy competition and yet prevent an excess of subsidized tonnage at low prices that foster destructive competition. The trigger-price mechanism installed in 1978 was reasonably successful for about two years due in part to the improved market conditions in 1978 and

1979, as well as early 1980. Subsequently when the market in Western Europe and the United States deteriorated, the trigger price was circumvented and then widely ignored, as steel came into the United States during late 1981 at prices ranging from $50 to $70 per ton below those stipulated by the trigger-price mechanism.

The result was a rapid deterioration of the U.S. price structure, which brought on heavy losses for the domestic steel companies, as well as those companies exporting to the United States.

The Western Europeans justified their actions with respect to the trigger price by maintaining that U.S. producers were selling steel below trigger prices, and this made it impossible for them to enter the U.S. market. They argued further that the devaluation of their currencies made it possible for them to produce steel in terms of U.S. dollars at costs below the trigger price. Under these circumstances, a number of companies sought a preclearance to ship steel into the United States below trigger prices. When this was denied, they proceeded to ship steel in at prices well below trigger, which brought on an investigation by the Commerce Department to determine whether or not steel was being dumped in the American market.

The logic employed by some Western European producers and a number of others who joined in the effort was that the U.S. industry in the first half of 1981 was profitable and operating at a high rate of capacity; therefore, this lower-priced steel would not cause injury. The orders at prices well below trigger were taken in the spring of 1981 with the U.S. industry operating above 85 percent of capacity and the importers, as well as the Western European producers, calculated that with this high rate of operation, the additional imports at lower prices would not cause a disturbance. Unfortunately this logic did not pay off since the American market dropped sharply in the third and fourth quarters thereby cutting steel operations to less than 60 percent of capacity.

It was precisely during this declining market that imports came in in very large tonnages. In August 1981, the figure was 2.02 million tons, an all-time record for any month and this constituted 24.6 percent of the U.S. market.[25] Much the same was true in November of that year, as steel orders continued to slide and imports remained at high levels, constituting 25.8 percent of the market. As a consequence, dumping suits were filed by the U.S. producers to cut the degree of penetration of low-priced subsidized imports into the United States.

Import penetration, if it extends much beyond 15 percent of the U.S. market (and it is beyond that level now), will result in the closure of more steel facilities, reducing the size of the industry, so that the economy in the not-too-distant future will have to depend on high-priced imports. What remains of the U.S industry after these closures should be able to operate profitably. However, the adjustments will serve to demonstrate the risks to steel investment and are likely to pose strong disincentives to future capacity installation at sharply inflated capital costs. In the final analysis, if the solution to the import problem is not developed, steel users will pay higher prices as the U.S. steel industry shrinks.

The solution must be aimed at restraining subsidized imports that are sold at prices substantially below their cost of production. Steel imports, which have most often sold at prices that cover their costs of production, present fair competition. However, since the nationalization of a number of steel companies, the drive to maintain employment in depressed times (which has led to subsidies) has changed the picture dramatically. The solution to the problem must involve some tonnage limitation accompanied by stipulated price levels, such as the EEC uses in limiting its steel imports. A mechanism such as the trigger price failed because it was too easily circumvented. The combination of the tonnage limitation and stipulated prices has worked successfully in limiting imports to the EEC and should be tried in the United States. Any solution should allow imports that comply with fair-trade principles but subsidized steel imports should not be permitted to injure the domestic industry.

Discussions of imports often involve a cost comparison between the U.S. industry and those of other countries. During the past 10 years, these comparisons have varied due to a number of factors not the least of which is the exchange rate between currencies. In 1977, a study by the Council of Wage and Price Stability indicated that the European costs were only slightly below those in the United States and the freight, which had to be added, made European steel noncompetitive in the American market if its prices covered the full cost of production. The Japanese costs, on the other hand, were low enough so that with the freight, their products were competitive in the American market. Recent changes in the relative value of currencies have altered this position particularly with regard to the Europeans so that their costs are lower in terms of dollars than they were in 1979 and early 1980. However, in reality, their costs have been increasing in relation to previous years as labor, freight, energy, and raw materials have advanced. The recently made wage comparisons indicate that European wages, in terms of dollars, have declined in the last few years. This again is a function of the exchange rate since the comparison is made in terms of dollars.

There is a substantial difference between the wage rates paid by the steel industry in the United States and those of some other countries. This is particularly true of Japan, Taiwan, and South Korea. The difference between the U.S. and some Western European country wage rates is not as great. In 1981, with currency exchanges taken into account, the U.S. wage rate on an hourly basis was $9 more than the Japanese rate. In terms of Taiwan and South Korea, where the wage rate is less than $4 per hour, the difference is more than $15. This differential places the industry in the United States at a distinct disadvantage, which could be reduced by increased productivity.

On the bright side, the U.S. steel industry could experience a significant growth in demand if the energy programs, which have been discussed at great length, are put into operation. Much has been said about making the United States self-sufficient in terms of energy and reducing its dependence on imported oil to a minimum. In order to accomplish this, the coal reserves in the United States would have to be utilized to a much higher degree. Coal could be processed into oil and gas to fill a share of the needs of the economy, which is now taken care of principally by oil imports. If this program is to accomplish its objective, it will be necessary to mine 500 million additional tons of coal above the present production level of 800 million tons. This will require opening new mines as well as an increase in railroad hopper cars to carry the coal to gasification plants, which, in themselves, will require large amounts of steel for their construction. Further, once the coal is gasified and liquefied, additional pipelines will be required to take these products to their destinations. Thus, the amount of steel required to mine the extra 500 million tons of coal, transport it to the processing plants, build the plants and then carry the products to market by pipeline will be large indeed and could more than compensate for the amount of steel lost in automobile production as cars are downsized and imports take a larger share of the market.

The basic problem here is the willingness of the administration to undertake and promote coal liquefication and gasification on the scale necessary to meet the future energy needs of the country. At present, there seems to be a lack of enthusiasm, due to the temporary glut of oil, and price decreases (although exorbitant by any standard of the last decade) in the past months. This lack of enthusiasm is unfortunate since the development of our coal reserves to provide energy is a long-run necessity and should be undertaken soon.

The long-term market outlook for steel is also brightened by the need to provide for a renewal of the nation's infrastructure, which has, in many respects, become woefully obsolete. Thus extensive replacement is an absolute necessity. Starting in the mid-1960s the infrastructure has been left to deteriorate, as government-spending priorities have been shifted away from capital projects toward the funding of social-welfare programs. By the early 1970s, it had already become apparent that highways, bridges, tunnels, water-sewage treatment plants, dams, transportation systems, and other public facilities throughout the country were in serious need of repair and replacement.[26] Nevertheless, over the past ten years, although a variety of government agencies have catalogued case after case of worsening neglect, spending for new projects as well as for maintenance and repair has declined sharply to a fraction of its former level.

Measured in constant 1972 dollars, nonresidential investment by states and localities has declined from more than $20 billion annually to approximately $2 billion, leaving a constant-dollar gap, or spending backlog, that will exceed $180 billion by 1985, representing the current-dollar equivalent of some $500 billion.[27] This bill for long-term neglect of the nation's infrastructure cannot be paid unless a new set of priorities is adopted to channel massive resources into the rebuilding efforts. This will also create a large number of jobs. Meanwhile, there is an increasing threat to the public health and safety, industrial and residential growth is being curtailed, efficiency is being reduced throughout the economy, and a significant pent-up demand for steel is continuing to grow.

In 1982, the U.S. steel industry shares the fate of its counterparts throughout the world who are mired in a deep depression. As a consequence of the decline in operations and earnings, more U.S. capacity will be eliminated as some companies either shrink in size, close their doors, or merge with others. The U.S. steel companies that survive, if proper steps are taken in the years immediately ahead, should be profitable and could be competitive on a worldwide basis. To put themselves in such a position, these companies will have to consider seriously the elimination of obsolete facilities and high-cost operations. If these cannot be replaced, they should be abandoned so that what remains is fully modernized.

A program of shrinkage and modernization can be a keynote to profitability in the years ahead. Such a program should not result in too great an overall loss in capacity, for facilities should only be abandoned when their replacement is not practical from a technical or economic point of view. It is far better to abandon a plant or a part of a plant that a company could not afford to modernize rather than have it limp along draining profits from the rest of the operation. Further, the size of a company's production is no longer the criterion of its success, but rather efficiency and profitability should take precedence and, if it is necessary to reduce capacity to improve profitability, this should be done.

In the modernization of facilities, the latest technology should be employed. For example, the blast furnaces, some of which leave much to be desired, could benefit from technology that is available from West Germany or Japan. However, this does not mean the widespread construction of 10,000-ton-a-day blast furnaces since they are only practical in very large plants. What is much more important in most U.S. plants is the efficient operation of 3,000-ton and 5,000-ton furnaces. Further, in terms of finishing facilities, it would be well for the industry to examine its wide hot-strip mills, some of which were installed before World War II and in spite of renovations are noncompetitive with the recent second-generation strip mills installed throughout the United States in the 1960s and early 1970s.

Further the large integrated steel companies, if they have not already done so, should give attention to diversification into nonsteel areas that can be profitable. It is important that these ventures be selected for their quality and growth potential rather than the possibility of obtaining them at bargain prices. Very often bargains have been troublesome and in the long run proved very costly.

Diversification, which is profitable, is a means of strengthening a steel company and improving its capability. Additional revenues, which come from nonsteel operations, have been used and can be used to provide funds to improve and modernize steelmaking facilities, for which funds would not otherwise be available. This was particularly true in a number of years during the last decade when steel operations were not adequately profitable to generate cash for their own replacement.

The shrinkage, modernization and diversification approach to improve competitive position could contribute significantly to a long-range solution to the trade problem, provided heavily subsidized steel is not permitted into the country. No amount of modernization and reorganization will make it possible for a company to compete with steel that is sold at 15 percent to 20 percent below its total cost of production. The problem of destructive competition will not be solved by modernization alone. It will also require a rational approach to trade.

By 1985, the U.S. steel industry, with its capability to produce raw steel reduced to 127 million to 129 million tons and a plant that is 85 percent to 90 percent efficient, will be in a much better position to meet world competition.

Notes

1. American Iron and Steel Institute, *Annual Statistical Report for 1980,* p. 55, except for 1981 which is a preliminary figure issued by the institute.

2. American Iron and Steel Institute, *Annual Statistical Report for 1980,* p. 13.

3. Ibid.

4. Ibid.

5. Ibid., p. 10.

6. Figures obtained from executive personnel at the Cleveland Cliffs Iron Company, Cleveland, Ohio.

7. *Coal Statistics International,* no. 2 (New York: McGraw-Hill, February 1981):3.

8. William T. Hogan, S.J., and Frank T. Koelble, *Analysis of the U.S. Metallurgical Coke Industry* (Washington, D.C.: U.S. Department of Commerce), pp. i and ii.

9. Estimates provided by the Engineering Division of Koppers Company, Pittsburgh, Pennsylvania.

10. American Iron and Steel Institute, *Annual Statistical Report for 1980,* p. 55.

11. American Iron and Steel Institute, *Annual Statistical Report for 1945,* p. 29.

12. Ibid.

13. *The New York Times,* 6 January 1949, p. 4.

14. American Iron and Steel Institute, *Annual Statistical Report for 1977,* p. 53; and *Annual Statistical Report for 1978,* p. 55.

15. American Iron and Steel Institute, *Annual Statistical Report for 1978,* p. 55.

16. American Iron and Steel Institute, *Annual Statistical Report for 1979,* (Washington, D.C.: 1980), p. 55.

17. "Technology and Steel Industry Competitiveness," (Washington, D.C.: Office of Technology Assessment, Congress of the United States, 1980).

18. William T. Hogan, S.J., and Frank T. Koelble, *Direct Reduction as an Ironmaking Alternative* (Washington, D.C.: U.S. Department of Commerce, 1981), p. 23.

19. *Technology and Steel Industry Competitiveness,* (Washington, D.C.: Office of Technology Assessment, Congress of the United States, U.S. Government Printing Office, June 1980), p. 87.

20. David H. Clark, "Computer Process Control in the Steel Industry" (Paper presented at International Iron and Steel Meeting, Toronto, Canada, October 1981.)

21. U.S. Congress, House of Representatives, "Message from the President of the United States Relative to Our Federal Tax System," 87th Congress, 1st sess., 20 April 1961, House Document No. 140, p. 5.

22. United States Steel Corporation, *Annual Report* (Pittsburgh 1973), p. 11.

23. American Iron and Steel Institute, *Weekly Raw Steel Production* (March 30, 1981, and March 15, 1982).

24. Figures obtained from the United States Steel Corporation's Commercial Research Department.

25. American Metal Market, September 30, 1981, p. 1.

26. William T. Hogan, *The 1970s: Critical Years for Steel,* Lexington, Massachusetts, D.C. Heath and Company, Lexington Books, 1972, pp. 95-100.

27. Morgan Guaranty Trust Company, *Morgan Guaranty Survey,* New York, July 1982, pp. 11-15.

6

Prospects for Canada, Australia, South Africa, and New Zealand

Canada

The Canadian steel industry was an exception to the downward production trend in the industrialized world after 1974. In 1974 it reached an all-time record of 13.6 million tons of raw-steel production, which declined slightly in the following year as production fell by approximately 4.5 percent to 13.0 million tons. In 1976 it recovered to 13.3 million, and by 1977, which was a poor year for the rest of the industrialized world, Canadian production surpassed its 1974 peak. In the following two years it surged ahead, achieving a record 16.1 million tons in 1979.[1] Canada was the only country in the industrialized world to surpass its 1974 record by a substantial margin, not only in physical output but also in profit performance.

The Canadian industry is currently one of the most efficient in the world, demonstrating an ability during three decades, in which its output increased from less than 3.0 million tons to over 16.0 million tons, to operate at high rates of capacity utilization, increase employment, and improve profitability. This outstanding performance, particularly in the past five years, differs markedly from that of most of its foreign counterparts, many of whom have been mired in a depression since 1975.

During the post–World War II period, the Canadian steel industry experienced a remarkable sixfold growth as raw-steel output rose from 2.6 million tons in 1947, to 16.1 million tons in 1979. A considerable portion of this growth took place between 1960 and 1979, as production moved from 5.2 million tons to the record 16.1 million tons.[2]

Canada's steel industry can be divided into two broad categories: the integrated producers and the small electric-furnace producers, which are often referred to as minimills. The first category lends itself to further classification by separating the large privately owned companies—Algoma, Dofasco, and Stelco—from the smaller, publicly owned, integrated companies—Sidbec-Dosco and Sydney Steel.

The privately owned Big Three, which over the past twenty-five years have represented between 75 and 80 percent of the industry's output, have been consistently profitable and are the key to the Canadian steel industry's success. Their earnings have ranged, in many recent years, from 7 percent to 9 percent of sales. Sidbec-Dosco and Sydney Steel, however, have had their problems due, in part, to a heavy debt structure involving large interest payments resulting in substantial losses. These have necessitated government subsidies to allow the companies to remain in business.

The second group of steel companies, referred to as minimills, produces steel in electric furnaces, using scrap as a furnace charge with a minimal amount of purchased DRI. For the most part, these companies are privately owned, although there is some government participation in two of them, Ipsco and Burlington. This category of producers also lends itself to a further grouping with Atlas Steels, a specialty-steel producer separated from the remainder. This company accounts for approximately 2 percent of the nation's total steel output.

Table 6–1 lists the steel producers in Canada according to classification.

The principal operations of Algoma, Dofasco, and Stelco are all located in the Province of Ontario, whence they serve the Canadian as well as the international market. Sidbec-Dosco and Sydney are located in Quebec and Nova Scotia respectively, from where they serve local as well as the international markets. The concentration of Canada's steel industry geographically is further pointed up by the fact that three of Canada's six minimills are also located in Ontario.

Again, unlike the rest of the industrialized world, the Canadian steel industry has expanded its facilities in the last few years. Stelco, the largest producer, has built a greenfield-site integrated plant on the north shore of Lake Erie. Currently, this plant's capacity is limited by a continuous caster that can process approximately 1.3 million tons; however, its blast furnace and basic-oxygen capabilities will permit it to produce 2.0 million tons. A continuous hot-strip mill is under construction at this site and will be in operation in late 1982. The cost of the mill to date is $1.3 billion. Dofasco is currently constructing a second hot-strip mill at a cost of $400 million, which will require more steel to maintain it in operation. Algoma is constructing a seamless-pipe mill for the production ultimately of 300,000 tons of oil-country goods. Lasco, one of the minimills, has recently doubled its capacity from approximately 500,000 to in excess of 1.0 million tons. Ipsco, an electric-furnace operation that is a significant producer of pipe, has recently increased its capacity by some 150,000 tons.

In terms of the next five years, there will be very little further increase in the Canadian steel industry's capacity, since most of the companies that have grown recently will need time to adjust to their new potential. The total capacity of the industry at the present time is between 19.0 million and 20.0 million tons.

One of the reasons for the success of the Canadian steel industry is that it has consistently maintained its capacity below the level of demand so that operations for many years have been at 90 percent or better of capacity. Further, it is a relatively new industry, since practically all of its facilities

Table 6-1
Steel Producers in Canada

Producers	Location
Integrated companies	
Algoma Steel Corp.	Sault Sainte Marie, Ontario
Dofasco, Inc.	Hamilton, Ontario
Sidbec-Dosco, Ltd.	Montreal, Quebec
Stelco, Inc.	Toronto, Ontario
Sydney Steel Corp.	Sydney, Nova Scotia
Nonintegrated companies	
Burlington Steel, Division of Slater Steel Industries, Ltd.	Hamilton, Ontario
Interprovincial Steel and Pipe Corp., Ltd.	Regina, Saskatchewan
Ivaco, Ltd.	Marieville, Quebec
Lasco	Whitby, Ontario
Manitoba Rolling Mills Ltd.	Selkirk, Manitoba
Western Canada Steel Ltd.	Vancouver, British Columbia
Specialty-steel company	
Atlas Steel	Welland, Ontario

Source: American Iron and Steel Institute, *Iron and Steel Works Directory of the United States and Canada* (Washington, D.C.: 1980).

now in operation have been installed since the close of World War II and most of them since 1960.

Raw Materials

In terms of raw materials, Canada has an abundant supply of iron ore that will last for many decades. These resources have been developed in several ways:

Individual Canadian companies have opened ore mines;

Groups of steel companies from Canada and the United States have participated in joint ventures;

The United States Steel Corporation has undertaken a major ore development in Canada.

The four outstanding developments in terms of iron-ore production included:

The formation of the Iron Ore Company of Canada that is composed of the following steel producers and ore companies with their percentage participation:[3]

The Hanna Mining Company	27.24%
Bethlehem Steel Corporation	19.41%
National Steel Corporation	18.20%
Armco Steel Corporation	6.06%
Jones & Laughlin Steel Corporation	6.06%
Republic Steel Corporation	6.06%
Wheeling-Pittsburgh Steel Corporation	4.86%
Hollinger Mines Ltd.	8.27%
Labrador Mining & Exploration Co. Ltd.	3.84%

Second, the Wabush Mines, which is also composed of steel and ore companies, has the following eight participants:

Steel Company of Canada	25.6%
Dominion Foundries & Steel Ltd.	16.4%
Youngstown Sheet and Tube Company	15.6%
Inland Steel Company	10.2%
Wheeling-Pittsburgh Steel Corporation	10.2%
Interlake, Inc.	10.2%
Finsider	6.6%
Picands-Mather & Company	5.1%

Third, the Quebec Cartier Mining Company founded by U.S. Steel Corporation to mine and concentrate ore.

Fourth, the most-recent development, Sidbec Normines, was formally established in 1976. This was a joint venture involving Sidbec-Dosco with 50.1 percent, BSC with 41.67 percent, and U.S. Steel with 8.23 percent. This group established mining operations at Fire Lake in Quebec and a concentrator and pellet plants at Port Cartier. The pellet plants have a capacity for 6 million tons, 3 million of which are specialized for use in Sidbec-Dosco's direct reduction plants.

With regard to coal, the Canadian steel companies import most of their coking coal from the United States where they have ownership interests in several mines.

Technology

In 1980, 55 percent of Canadian steel production was accounted for by the oxygen converter, 25 percent came from electric furnaces, and 20 percent from open hearths.[4] Open-hearth furnaces are in operation at the Hamilton Works of Stelco and the Nova Scotia plant of Sydney Steel. The Hamilton furnaces are perhaps the most modern open hearths in North America. There are five large units, all using oxygen. One has a capacity to produce 500 tons per heat and each of the other four are capable of 315 tons per heat. Stelco has no plans to replace these facilities since they have proven to be efficient and competitive. The open-hearth furnaces at Sydney Steel are old and most probably will be replaced in the near future.

The new Stelco plant is based on oxygen steelmaking and this will increase the percentage accounted for by this process.

The Canadian industry used continuous casting to produce 1.7 million tons (13.3 percent) of its output in 1975. By 1980, continuous casting accounted for 4.1 million (25.6 percent) of the industry's output.[5] Here again, the new Stelco plant with its continuous-casting unit and the unit recently installed at Lasco will raise the share of Canada's continuously cast steel to more than 30 percent within the next two years.

Sidbec-Dosco has installed two Midrex units for the production of DRI with a total capacity of 1.2 million tons. One of these is the most modern unit operating today and has achieved excellent results. It was originally rated at 600,000 tons per year but has operated consistently at an 800,000 ton level.

Stelco has an SLRN direct-reduction unit with a rated capacity of approximately 400,000 tons; however, the price of scrap has been such in the past few years that it was uneconomical to operate this facility.

In terms of the future, it is possible, with the proper price for gas, that at least one additional unit will be installed by Sidbec-Dosco to produce DRI for the trade.

Trade

During the past ten years, Canada has carried on a substantial trade in steel relative to the size of its industry. In the decade of the 1970s, it was a net importer of steel for the first five years and a net exporter for the last five years. Table 6-2 gives Canadian imports and exports of steel-mill products from 1971 through 1980. An examination of the table indicates why the Canadian industry was able to operate at a high rate of capacity during the entire decade. In 1975, 1976, and 1977, when the rest of the industrial world was experiencing problems, the Canadians reduced their imports considerably, and the domestic industry continued to operate at a high level. In 1976 and 1977, exports were accelerated, which also helps to account for the high operating rates in Canada during those years.

The principal market for exports is and will be the United States into which, in 1978, 1979, and 1980, a tonnage of 2.4 million was shipped in each year.[6] Because of their proximity to the U.S. market, the Canadian

Table 6-2
Canadian Imports and Exports of Steel Products: 1971-1980
(millions of metric tons)

Year	Imports	Exports
1971	2.1	1.5
1972	2.2	1.5
1973	2.1	1.6
1974	3.1	1.7
1975	1.6	1.3
1976	1.3	1.8
1977	1.5	2.2
1978	1.6	2.7
1979	2.2	2.6
1980	1.5	3.5

Source: American Iron and Steel Institute, *Annual Statistical Report* (Washington, D.C.: 1975), pp. 82 and 83.

mills were able to obtain preclearance in relation to the trigger-price mechanism and, in most instances, to ship their steel below trigger prices.

The 1965 treaty between Canada and the United States in respect to automobile imports and exports, which virtually created a common market for automobiles, accounts for a significant percentage of steel exports. Since Canada does not have a major stamping plant for the production of automobile-body parts, these are made in the United States, shipped to Canada, and assembled into automobiles, many of which are returned to the United States. Very often steel from Canada is stamped into the body parts that are returned to Canada to be assembled into automobiles.

The Canadian industry, despite its singular success at home, has prepared itself to compete with companies in other industrialized countries, as well as the Third World. It recognizes that it faces a different set of conditions in the 1980s, particularly in regard to world steel, than existed in the 1970s. The future will by no means be identical with the past, and consequently, the results of the last few decades can not be taken for granted in the years that lie ahead. It is virtually certain that the industry will grow somewhat during the remainder of the 1980s, however, the pace will be slower than it has been. There are also definite possibilities that the smooth operation that the industry has enjoyed may be somewhat disrupted from time to time due to worldwide steel conditions.

The Canadian companies, particularly the big three, have made wise management decisions regarding capital investment, as well as operating and sales practices. They have been successful despite the fact that they have no particular advantages in terms of raw materials. Virtually all of the coal comes from the United States at a price equal to and in some instances slightly higher than that paid by the U.S. steel industry. A significant portion of iron ore is imported, although large quantities are now obtained from domestic sources. These sources are located in Labrador and Quebec and, as previously indicated, involve partnerships with a number of U.S. companies.

Government Attitude and Assistance

The Canadians have benefited from the atmosphere created, in part, by their government, which has helped them to function profitably. Depreciation, environmental, and tax policies of both the federal and provincial governments have been helpful. Further, the government has taken a reasonable attitude toward trade, particularly when imports have entered the market at low prices.

Substantial assistance has been given to Sydney Steel Corporation to help it reverse its record of losses, which are due, in large part, to its debt structure. In 1981 the provincial government of Nova Scotia assumed full responsibility for Sydney Steel's $300 million debt, thus relieving it of a substantial interest burden. The federal government agreed to provide $77 million toward a $96.2 million rehabilitation program with the provincial government paying the difference. The money will be used for repairs to the coke ovens, one blast furnace, and for updating the open-hearth furnaces and rolling mills.

The other publicly owned integrated steel company, Sidbec-Dosco, is also seeking help from the federal government to relieve it of a heavy debt. This was a result of a 50.1 percent participation that the company took in Sidbec-Normines, an iron-ore development of the Fire Lake Mine as well as the construction of facilities for concentration and pelletizing at Port Cartier. Total investment in the project amounted to $625 million. In addition to Sidbec, BSC and United States Steel also participated.

The project was expected to produce 6.0 million tons annually. At this rate of operation it would have been a profitable proposition. Unfortunately, it never operated at its full potential and has incurred substantial losses. Sidbec must pay interest, as well as amortization on its share of the debt and absorb half of the loss. In 1979, the interest, amortization, and share of losses came to a total of $20.8 million. Since that time, the situation has worsened, as losses continue to mount with the reduced operating rate at Fire Lake.

Currently Sidbec is seeking government aid in the hope that it will be relieved of a major share of its debt. As of mid-1982 this has not happened.

This appeal, as well as the recent development at Sydney Steel, point up the difference between publicly and privately owned companies in Canada. When in difficulty, the publicly owned companies can appeal to the government for a bailout. This is not possible for the privately owned companies; consequently, they have to be much more cautious about making large financial commitments if they are to remain viable.

The Canadian steel industry in the first half of 1982 suffered a 20 percent loss in production compared with the same period in 1981. However, the private-sector companies of the industry are in good financial condition and will be able to weather the storm. With their new facilities, they should emerge as strong units, fully able to compete in the world market.

Australia

Steel production in Australia is currently provided by one company—the Broken Hill Proprietary Company Ltd. (BHP). A number of finishing facilities in the country are provided by other companies. However, it is accurate to say that raw-steel production in Australia is equated with that of BHP. Sometime in late 1982 or early 1983, an independent minimill, currently under construction near Melbourne, will begin operations producing 150,000 to 200,000 tons of raw steel to be converted into small structurals, bars, and concrete reinforcing bars.

During the past two decades, raw-steel output has grown steadily. In 1960, it was approximately 3.75 million tons. By 1969, it had grown to 7.0 million tons, almost double the 1960 figure. During the 1970s, growth had been relatively slow, reaching the high point of 8.1 million tons in 1979.[7]

BHP operates three fully integrated plants. The largest is the Port Kembla Works, south of Sydney, that has an annual capacity of 5.5 million tons. The second largest is the Newcastle Works at Newcastle, north of Sydney, with an annual capacity of 2.7 million tons, and the third is the plant at Whyalla, near Adelaide, with a 1.3-million-ton capacity. The total for the company is 9.5 million tons.[8] In addition to the three integrated steel plants, BHP operates a blast furnace that produces iron for the trade; it is at Kwinana in West Australia.

BHP is a complex operation of which steel is but a part. It operates a mineral division that supplies iron ore and coal not only to its own plants but also to other countries around the world. In addition, there is an aluminum division and an oil and gas division that has been quite profitable.

Most of the steel produced by BHP is consumed in Australia. However, the company has taken advantage of export opportunities from time to time. Interestingly enough, imports involved considerable tonnage during several years in the past decade, indicating that during some years BHP's

Table 6-3
Exports and Imports of Australian Steel Products: 1970-1979
(millions of metric tons)

Year	Exports	Imports
1970	0.753	0.49
1971	0.538	1.52
1972	1.292	2.12
1973	1.314	0.76
1974	1.281	1.48
1975	2.142	0.93
1976	2.361	1.40
1977	2.483	0.10
1978	2.571	0.60
1979	1.707	0.50

Source: International Iron and Steel Institute, *Steel Statistical Yearbook 1981* (Brussels), pp. 22 and 24.

capacity was not adequate to supply Australian demands. This situation with some minor exceptions, has changed. Table 6-3 lists exports and imports of steel products for Australia from 1970 to 1979.

As recently as October 1981, BHP had plans to increase its capacity by a small amount with improvements at the Port Kembla and Newcastle Works. There was need for additional ironmaking facilities at Newcastle, and serious discussions were held concerning the prospect of building a new blast furnace at that location to replace some of the old units as well as to furnish additional iron output. However, because of the decline in business, the project was shelved.

As steel conditions worsened throughout the world and imports came into Australia in increasing tonnages at low prices, BHP had to reassess its long-term plans. The result was a decision to close down a number of facilities and improve the efficiency of the remaining capacity. Some three blast furnaces are due for permanent closure, including the furnace at Kwinana Works, as well as one at Whyalla and Newcastle. The furnace at Newcastle is a small operation whose origin goes back many years. The production lost from the furnace to be closed at Whyalla will be compensated for by enlarging the other furnace at the plant.

One of the most radical changes is the elimination of the open-hearth facilities at Port Kembla, which will reduce the capacity of BHP by approximately 750,000 tons. These decisions were made, in great part, because of the increase in steelmaking costs while productivity remained stable. BHP is a privately owned company and, therefore, receives no subsidies from the government and must operate efficiently if it is to remain profitable. Thus, marginal facilities have to be eliminated.

In terms of the future, BHP will concentrate its capital expenditures on modernization and improving facilities. Continuous-casting units will be installed to raise the amount of steel thus treated from the current 13 percent to over 30 percent. Further, the elimination of the open-hearth shop will raise the percentage of steel made by the oxygen process to more than 90 percent.

In some areas of production, BHP has highly efficient modern units. For example, in the mid 1970s, John Lysaght Ltd., which is partly owned by BHP, built a new thoroughly modern 82-inch-wide hot-strip mill. It was commissioned in 1978 with an ultimate annual capacity of 5 million tons. This mill has a coil box installed that contributes significantly to its efficiency and has enough capacity to meet the needs of BHP and Australia for sheets for some time to come.

Raw Materials

The Australian steel industry is most fortunate in terms of its raw materials. Coal is mined within a very short distance of the Port Kembla and Newcastle Works, and although high in ash content, it makes an adequate coke at a reasonable cost. Iron ore is available in abundance. Most of the deposits, however, are located in western Australia and must be transported to the two plants on the east coast. The Whyalla plant has a more than adequate supply of iron ore located within a few miles. This deposit supplied the Australian steel industry until the middle 1960s when the vast new deposits of ore in the west were discovered and brought into production. Up until the early 1960s, before the ore discoveries in the west, there was an embargo on iron-ore exports since the known deposits were deemed necessary to supply to domestic industry.

Australia's greatest contribution to the world-steel industry in the last decade has consisted of supplying the world with abundant, rich iron ore, as well as large amounts of coking coal. Since the embargo on iron-ore exports was lifted, the country has become a highly significant part of the world's international trade in iron ore. Table 6–4 gives iron-ore exports from Australia from 1967 through 1980.

Australia's principal customer for iron ore is Japan. However, in the past ten years, a trade has been built up with a number of Western European countries that reached a peak of almost 14.5 million tons for the EEC in 1976. Since then, this trade has declined somewhat. Coal is shipped in such large tonnages to Japan that Australia has replaced the United States as Japan's leading supplier. In 1980, about 60 million tons of ore and 24 million tons of coal were supplied to the Japanese steel industry.

Table 6-4
Exports of Australian Iron Ore: 1967-1980
(millions of metric tons)

Year	Exports	Year	Exports
1967	9.2	1974	83.6
1968	16.3	1975	80.4
1969	26.9	1976	81.4
1970	44.6	1977	78.9
1971	52.0	1978	75.3
1972	58.1	1979	78.4
1973	74.2	1980	80.1

Source: International Iron and Steel Institute, *Steel Statistical Yearbook 1981* (Brussels).

In 1980, Australia had a relatively small percentage—some 10.3 percent, or 800,000 tons—of its steel continuously cast. This will increase significantly in the years ahead. In the same year, 75.7 percent of Australian steel was produced by the basic-oxygen process, with 21.5 percent from the open hearth, and a small 2.7 percent from the electric-furnace process.[9] These figures will change significantly in the decade ahead as much of the open-hearth capacity is taken out of service. Improvements in blast-furnace production will be brought about by increasing the size of one furnace at Whyalla, as well as the number-five furnace at Port Kembla.

Because Australia had such an abundance of steelmaking raw materials, a plan was proposed in 1968 to build a large steel mill using the raw materials and shipping semifinished steel. It was hoped that such a plant, with a number of steel companies as partners, would ultimately produce 10 million tons of semifinished steel to be distributed among those engaged in the project. Unfortunately, the plan never materialized.

Another joint venture was conceived by two U.S. companies, Armco and Kaiser, that was to have resulted in the construction of a large integrated plant at Jervis Bay, about 100 miles south of Sydney. A large tract of land was selected, but the plan was abandoned when it was not possible to fill the government's requirement that an Australian partner be found to take 50 percent interest. The third joint venture with thirteen partners was discussed in the early-to-mid 1970s. This project involved U.S., European, and Japanese companies, but proved to be too unwieldy and came to naught.

The government of the Western Australian Province has been attempting to develop a joint venture for steel production for a number of years and will continue to do so. It has been stipulated in its contracts with the ore producers that they do secondary processing and also study the feasibility of a steel mill. The participants in the Mount Newman Project agreed to install secondary processing equipment and by 1988 to present a feasibility study for a one-million ton steel plant.

A principal factor in financing the development of iron-ore deposits in Australia was the long-term contracts for large tonnages of ore with the Japanese steel industry. A number of locations have been developed, while others are known and awaiting development. Major ore deposits exist, with an estimated total reserve of about 30 billion tons, in some eighteen locations throughout the country.

Australia and Brazil have been vying for the position of number-one exporter of iron ore to the world for several years and undoubtedly will continue to do so. In 1980, exports from each of these countries were approximately 80 million tons.

Australia will continue to supply raw materials to much of the world's steel industry. BHP will remain diversified with no planned increase in steel output during the remainder of the decade. The company will export from time to time; however, this will by no means represent a major share of its activity.

South Africa

The South African iron and steel industry consists of a number of companies, the largest of which by far is the South African Iron and Steel Industrial Corporation Ltd., known as ISCOR, having a raw-steel capacity of 7.5 million tons. The second largest company is Highveld Steel and Vanadium Corporation, with a capacity of approximately 900,000 tons. There are several smaller electric-furnace operations including Dunswart Iron and Steel Works with an annual capacity of approximately 250,000 tons, Scaw Metals Ltd., and the Southern Cross Steel Company, as well as some others that operate electric furnaces with annual capacities of less than 300,000 tons.

In 1970, South African raw-steel production was 4.8 million tons. In 1974, a most prosperous year for the world steel industry, South African production rose to 5.8 million tons, of which ISCOR produced over 4 million tons. Unlike the countries in the industrialized world, South African steel production increased steadily during the second half of the 1970 decade, rising to 9.1 million tons by 1980.[10] Much of this increase was due to the growth of ISCOR, which continued to represent the lion's share of South African steel production. ISCOR is a relatively young company, having been established by statute in 1928 and having produced its first steel in 1934 at the Pretoria Works. It currently operates three integrated plants

with blast furnaces, basic-oxygen facilities, and with the exception of Newcastle, electric furnaces. The largest of its three integrated plants is its Vanderbijlpark Works. The newest facility is the Newcastle Works, which was
built in the early to mid-1970s.

During the next few years, ISCOR has no plans to expand its raw-steelmaking capacity beyond the current 7.5 million tons. The company is constructing 600,000 tons of DRI capacity based on coal. Further, because of
rather poor coking coal, it will install a Japanese process for briquetting a
portion of its coal to be charged into the coke ovens. These two improvements should give the company more ironmaking units and better coke to
improve its blast-furnace productivity. ISCOR has a large comprehensive
capital program involving improvements to its coke ovens, furnaces, and
rolling facilities at all three plants. The current plan is long range in nature,
extending through the remainder of the decade. As carried out, the program
will enable the company to continue to produce a wide range of products including plate, sheets, tinplate, structural sections, rails, wire rods, and wire
products.

Highveld Steel and Vanadium Corporation, a privately owned company, is a producer of steel, but it has a substantial interest in the production of vanadium. It operates both electric and oxygen steelmaking furnaces
with a total capacity of 900,000 tons, and it plans to increase its current raw-
steel capacity to 1.2 million tons in the next few years.[11]

One of the smaller companies, Dunswart, operates a successful direct-
reduction unit based on coal. It expects to add other units. This could either
increase the capacity or maintain the same capacity with less reliance on
scrap, which is in short supply in South Africa.

In 1973, ISCOR inaugurated two major projects. One known as the
Sishen-Saldanha Bay project involved extensive developments at the Sishen
ore mine to produce between 15 million and 20 million tons a year, the construction of a railroad of 860 kilometers, or 570 miles, from the mine to
Saldanha Bay, where ore-loading facilities capable of handling 15 million to
20 million tons a year were to be constructed. The project was termed the
largest development scheme ever undertaken in the Republic of South
Africa. Construction was started in 1973 and finished in 1976. Increasing
amounts of ore were exported in subsequent years with the tonnage reaching 14.5 million for the 1980 fiscal year.[12]

The second project was also a grand scheme, however, it came to
naught. It involved the possibility of constructing a greenfield-site steel mill
at Saldanha Bay to produce, in its first phase, 3 million tons of semifinished
steel. It was to be a joint venture involving several European and possibly
one American partner. ISCOR was to hold 51 percent of the shares with
Voest of Austria as the principal partner with 26 percent of the shares. The
plant was to have two blast furnaces, three 160-ton basic oxygen converters,

and three continuous casting units. In subsequent phases, it was to be extended to 12 million tons. Unfortunately the steel depression, which began in 1975 in most parts of the world, had a sobering effect on the prospective partners, and in 1976, the plan was quietly abandoned.

South Africa has an abundance of rich iron ore. However, its coal, although plentiful, leaves much to be desired in terms of coking quality. To solve the problem that this poses, ISCOR, as previously mentioned, will adopt the Japanese process for briquetting coal in the production of coke.

In terms of technology, South Africa produces approximately 65 percent of its steel by the oxygen process, 25 percent by the electric furnace, with the remaining 10 percent divided three to one in favor of the open hearth versus the Thomas process. Continuous casting has increased from approximately 20 percent of the steel produced in 1975 to almost 52 percent in 1980.[13]

South Africa has become a factor in exporting steel products during the past decade. In 1972, exports surpassed 500,000 tons and, by 1976, they reached 1 million tons. In 1978, a record 2.2 million tons were exported, a volume that has been sustained through 1980. Relatively little steel is imported. During the remainder of the 1980s, South Africa will continue to supply large amounts of iron ore to steel plants throughout the world from its Sishen—Saldanha Bay complex.

ISCOR, a semi-state-owned company, in which the government holds the A and B redeemable preference shares and private shareholders have the cumulative preference shares, will concentrate on improving its current facilities with investments at all three of its integrated plants.

New Zealand

New Zealand has a small steel industry based in great part on local iron sands that are available in large quantities. The use of this raw material began in the early 1970s when it was successfully turned into direct-reduced iron (DRI) using a rotary-kiln process. Prior to that time, a small amount of steel, 45,000 tons in 1966, was produced in the electric-arc furnace from scrap. After the introduction of DRI, steel output grew to more than 200,000 tons. The production was based on a combination of scrap and DRI with a three-to-one ratio in favor of the latter. The steel produced in this manner is continuously cast into billets for processing into rods, merchant bar, reinforcing bar, and small structural sections.

New Zealand imports both hot- and cold-rolled sheets, which are processed into pipe and galvanized sheets.

There are ambitious plans to increase steel production to about 800,000 tons based on the iron sands. In addition to the projected increase in raw steel, there are further plans to install mills to produce hot- and cold-rolled sheets. In 1980, raw-steel output in New Zealand stood at 230,000 tons.

Notes

1. International Iron and Steel Institute, *Steel Statistical Yearbook 1981* (Brussels 1981), p. 2.

2. Ibid., American Iron and Steel Institute, *Annual Statistical Report for 1950*, p. 94; and American Iron and Steel Institute, *Annual Statistical Report for 1960*, p. 130.

3. *Mineral Bulletin—MR148—Iron* Ore, Energy, Mines and Resources Canada, pp. 2 and 3.

4. IISI, *Steel Statistical Yearbook 1981*, p. 5.

5. Ibid., p. 7.

6. American Iron and Steel Institute, *Annual Statistical Report for 1980* Washington, D.C., p. 49.

7. American Iron and Steel Institute, *Annual Statistical Report for 1960*, p. 137; and International Iron and Steel Institute, *Steel Statistical Yearbook 1981*, p. 3.

8. *BHP Pocketbook* published by Public Affairs Department of Broken Hill Proprietary Co. Ltd. (Melbourne: January 1977), pp. 6–13.

9. IISI, *Steel Statistical Yearbook 1981* (Brussels), pp. 5 and 7.

10. Ibid.

11. Interview with executive personnel of Highveld Steel and Vanadium Corporation.

12. Iscor, 1980 *Annual Report* (Pretoria: South African Iron and Steel Industrial Corporation Ltd., 1980), p. 5.

13. IISI, *Steel Statistical Yearbook 1981*, pp. 5 and 7.

7

The Third World Moves toward Self-Sufficiency

The Third World, unlike the industrialized countries, has embarked on an ambitious steel-expansion program. Much of this was accomplished during the last half of the 1970s when virtually all of the industrialized world's steel industry was in a period of depression and contraction. The program stemmed in part from the second general conference of the United Nations Industrial Organization, which was held in March 1975 in Lima, Peru. The principal action of the conference was the adoption of the Lima Declaration on Industrial Development and Cooperation that set goals for the future industrialization of the developing countries. The declaration points out "that the developing countries constitute 70 percent of the world's population and generate less than 7 percent of the industrial production."[1] To remedy this situation, it states that the developing countries should endeavor to increase their industrial production "to at least 25 percent of total world production by the year 2000, while making every effort to ensure that the industrial growth so achieved is distributed among the developing countries as evenly as possible."[2]

This program required an increase in the industrial growth rate of more than 8 percent a year. It was also recognized that a 25 percent share could not be achieved in every industry. To compensate for this, suggestions have been made since the Lima conference that a higher percentage should be attained in some industries to bring the average to 25 percent. One of these industries is steel, and some proponents of the idea feel that the developing countries should produce 30 percent of the world's steel output by the year 2000. What this production will amount to in terms of actual tonnage obviously depends on total world steel output in 2000. Estimates vary widely. For example, one source puts the tonnage as low as 1.4 billion tons, while another, more-optimistic source predicts 1.7 billion tons. Thus, the target for the developing countries would be between 350 million tons if 25 percent of the lower number is accepted and 525 million tons if 30 percent of the higher figure is taken.

Since 1975, in the face of a worldwide depression in steel, the Third World countries increased their steel output by 67 percent, from a total of 60.3 million tons in 1974 to 100.8 million tons in 1980. In Latin America, aggregate output rose during the same period from 17.7 million tons to 29.1 million tons, an increase of 65 percent.[3] Other countries with significant

growth in their steel capacities and production include China, India, South
Korea, and Taiwan. Companies in these countries have all expanded during
the past five years and presently have plans for further expansion. Table 7-1
gives the output for the principal steel producers in the Third World in
1970, 1974, and 1980.

The total growth for the Third World from 1970 through 1980 represented an increase of almost 150 percent, as output rose from 42.3 million
tons in the former year to 100.8 million tons at the end of the decade. In
1981, it declined slightly to 99.8 million tons.

The objectives of the Third World countries are to develop as much
steel capacity as can be financed in order to supply their own needs and to
have some additional tonnage available for export, thereby improving their
balance-of-payments position. During the last five years, almost all of the
Third World countries capable of producing 2 million tons had plans for
expansion. Some of these plans have been carried out, while others have
been severely modified. Unquestionably, the 58-million-ton increase in its
steel output since 1970 and, more important, the 40-million-ton increase
since 1974 have eliminated market outlets for industrialized-world production, thereby aggravating trade relationships between and among the United
States, Western Europe, and Japan.

A country-by-country review of past achievements and future plans
indicates the course the Third World will follow during the remainder of the
decade.

China

The steel industry in mainland China has gone through a number of drastic
changes since World War II.[4] The first available production statistics that
are reasonably reliable date back to 1949 when some 158,000 tons of raw
steel were produced. This amount increased through 1954 when the 2-mil-
lion-mark was passed for the first time. For the next six years, with the aid
of Soviet technicians, the output of the Chinese steel industry increased
rapidly, reaching a total of 18.7 million tons in 1960.[5]

Because of political differences, the Soviets withdrew their assistance,
and steel production in 1961 dropped to 8 million tons. It came back slowly
and, by 1970, had reached 17.8 million tons.[6] A few large plants were
operating, but a significant amount of iron production in the early 1960s
came from small, so-called backyard furnaces that were producing a few
tons per day. Many of them with less than 300 tons of capacity per year
were abandoned in the mid- to late 1960s.

In the last ten years, the Chinese have made significant advances,
increasing production from the 17.8-million-ton figure of 1970 to 25 million
tons in 1975. In 1976, production dropped to 21 million tons, but since that

Table 7-1
Raw-Steel Output for the Leading Third World Producers: 1970, 1974, and 1980
(millions of metric tons)

Country	1970	1974	1980
China	18.0	26.0	37.0
Brazil	5.4	7.5	15.3
India	6.3	7.1	9.4
South Korea	0.5	1.9	8.6
Mexico	3.9	5.1	7.1
North Korea	2.2	3.2	5.8
Taiwan	0.4	0.9	4.2
Argentina	1.8	2.4	2.7
Venezuela	0.9	1.0	1.8

Source: International Iron and Steel Institute, *Steel Statistical Yearbook 1981* (Brussels), pp. 2 and 3.

time, output has increased rapidly, reaching 37 million tons in 1980.[7] In 1981, it dropped slightly to 35.6 million tons.

In 1978, with output at 31.8 million tons, the Chinese announced a program to raise steel production to 60 million tons by 1985 through the construction of ten iron and steel complexes to be built by the mid-1980s. Further, plans called for production to reach 160 million tons by the year 2000. A most ambitious program was laid out in which the state announced that it would build and develop a hundred projects between 1978 and 1985, including ten iron and steel complexes, eight coal mines, ten oil and gas fields, thirty power stations, six trunk railways, and five key harbors.

To facilitate this project, a $20-billion, eight-year, long-term trade agreement was signed with Japan in February 1978. The agreement was signed for Japan by Yoshihiro Inayama, chairman of Nippon Steel, and provided an opportunity for the Japanese steel industry to turn to a new market for steel and steel-technology exports. Nippon Steel agreed to construct a greenfield, fully integrated steelworks at Shanghai that would ultimately produce 6 million tons. The agreement also called for Chinese exports of raw materials including crude oil, coking coal, and steam coal to Japan. Coking-coal exports in 1978 were to be 300,000 tons, increasing to 500,000 in 1979, 1 million in 1980, 1.5 million in 1981, and 2 million in 1982. During the same period, the Japanese planned to export $7–$8 billion worth of plant and machinery and $2–$3 billion worth of construction materials to China.[8]

The steel-construction projects were awesome, since it was estimated that to increase production to 60 million tons by 1985, China would have to add 5 million tons of new capacity each year. Some observers doubted that this could be achieved. The Japanese, however, who were engaged in a considerable portion of this expansion, were quite confident that the objective could be reached. The program would require the improvement and expansion of four existing plants and the construction of three or four fully integrated greenfield-site works.

Part of the expansion of existing plants took place at Wuhan Iron and Steel Corporation, one of China's four major steel complexes. This plant, with an annual raw-steel capacity of 2 million tons, was constructed originally by the Soviets, and the Japanese, through Nippon Steel Corporation, added a wide hot-strip mill that was completed in 1978. In connection with the strip mill, three hundred Chinese were sent to Nippon Steel plants for training, and about the same number of Nippon Steel personnel was sent to Wuhan's plant to provide on-job training at the new mill.

The most significant project, which was also the first to be approved in the new program, was the Baoshan Steel Works near Shanghai. In May 1978, Nippon Steel signed a letter of intent with the Chinese government authorities, outlining the scope of responsibilities that Nippon Steel had agreed to undertake. These involved engineering, construction, and the initial operation of the 6-million-ton integrated plant at a site called Baoshan on the Yangtze River.

The plant, to be built in two stages, was to consist ultimately of two large blast furnaces, each capable of 9,000–10,000 tons of iron production per day, with supporting coke ovens and sinter plants. Steel-making equipment would ultimately consist of three 300-ton basic-oxygen converters, with part of the output continuously cast and part poured into conventional ingots and rolled down on a blooming-slabbing mill. Finishing facilities would consist of a hot-strip mill, a cold-strip mill, and a seamless-pipe mill. In addition to the products from these facilities, a substantial tonnage of semifinished steel was to be produced to feed other plants.

The first phase, providing 3 million tons, was to consist of one blast furnace, two basic-oxygen converters, a blooming mill, and a seamless-pipe mill. The second phase, with 3 million more tons, was to include the second blast furnace, the third oxygen converter, a continuous-casting unit, a hot-strip mill, and a cold-strip mill. The first 3 million tons of production were to provide 500,000 tons of blooms for the seamless-pipe mill, 1.5 million tons of billets to supply rerollers in the area, and 1 million tons of slabs for the Wuhan Steel Works hot-strip mill.[9] Nippon contracted to supply all the facilities, with the exception of the seamless mill that was assigned to Mannesmann of West Germany.

A second fully integrated plant, Chitung Steel Works, was proposed for the Hopeh Province, in northeastern China, with an ultimate capacity of

10 million tons. A West German consortium, consisting of Thyssen and several suppliers as well as the Dresdner Bank, was formed to bid on the project, the total cost of which was estimated at $15 billion. As a means of financing this project, China advanced the idea of compensatory trade by which it would sell copper, aluminum, zinc, lead, chromium, and molybdenum to West Germany to generate income with which to pay for the plant and equipment, as well as the know-how. The Germans preferred acquiring the right to market these materials internationally.

The first stage was to consist of 6 million tons of steel-making capacity. The facilities were composed of two blast furnaces, three 300-ton oxygen converters, a blooming mill, a continuous-casting unit, a hot-strip mill, and a cold-strip mill, as well as facilities for tinplate and galvanized sheets. In the second phase, three more blast furnaces would be added, a second basic-oxygen shop with two 300-ton converters, a second continuous-casting unit, a plate mill, a pipe mill, and a second cold-strip mill.

In addition to the two greenfield-site plants, the existing units, of which there were eight, each with 1 million tons or more capacity, were to be modernized and expanded.

The largest existing integrated plant is the Anshan Works that is currently capable of producing some 7 million tons of raw steel per year; this plant constitutes about 20 percent of the national output. The mill has eleven blast furnaces, most of which are small, but the most recent unit, installed in early 1978, is a 5,000-ton-per-day operation. Steel-making and finishing facilities consist of twenty-five open-hearth furnaces, two 150-ton basic-oxygen converters, and a variety of rolling mills including plate, hot-strip, rail, and pipe mills. Proposals were requested from several Japanese steel companies to modernize the plant and construct a second integrated works adjacent to it so that the ultimate production of the complex would be 15 million tons.

Several of the other existing integrated plants were due for significant expansion of from 3 million to 6 million tons. All five of the major Japanese steel producers were asked to participate in one or another phase of the expansion program.

The program also involved the development of iron-ore and coal facilities that would provide the raw materials to feed the expanded industry. In addition to developing the mines, several pellet plants were included in order to beneficiate the low-grade ore. Perhaps the largest raw-materials project was an iron-ore development project discussed with United States Steel Corporation, involving $1 billion.

The Chinese coal deposits, among the largest in the world with an output of as much as 600 million tons per year, were to be developed by U.S. companies that were asked to consider designing and constructing two surface mines, as well as five underground mines, and six coal-preparation plants.

In order to help operate the expanded Chinese steel industry, the Japanese agreed to train 2,000 Chinese technicians for the expansion of the Anshan Steel Works and 1,500 for the Baoshan Steel Mill.

To finance this vast expansion, the Chinese were offered billions of dollars of credit by the various countries that hoped to participate in the business. These loans had to be repaid at some point, and the Chinese were weighing the offers of the various suppliers very carefully, since the investment would be huge. In this respect, they looked for the best possible economic assistance. The Japanese had the inside track; however, it was quite certain that the Western European countries and the United States would be involved. This was all the more evident when, in 1978, the Chinese became aware of the fact that, due to the vast amount of construction they had requested, the Japanese could not handle all the work required.

The plants were so large and the expenses so great that governments had to become involved. Discussions were held as to the ability of the Chinese to repay all of the credits that had to be extended if not only the steel program but also others were to be carried out. Much of the debate revolved around the possibility that the Chinese might want to pay in the form of goods rather than cash. In fact, some firms withdrew from negotiations when the Chinese insisted that part of the repayment for capital investment be made in goods rather than cash.

The elaborate program that the Chinese had mapped out to modernize and expand their steel industry proved to be too ambitious. The same was true of programs for other basic industries they had hoped to develop in a very short time. In early 1979, it became apparent that industrial development on such a scale in so short a period was not practical, and it was necessary for the Chinese to retrench. At the Fifth National People's Conference, a revision of priorities was announced that put less emphasis on heavy industry and more on light industry and agriculture. One of the high officials announced that the medium-term, steel-production goal of 60 million tons by 1985 had been revised and lowered to 46 million tons.

This meant that the greenfield-site plant, which was projected at Chitung, had been indefinitely postponed, and there was much discussion about the remainder of the steel program. During the next year, it was announced that the second phase of the Baoshan steel mill, which would double the production to 6 million tons, was to be postponed. Much confusion reigned about the future of this plant, and it was finally settled that the second phase would be canceled although work would proceed on the first phase, resulting in the production of 3 million tons. Other ambitious projects like the construction of a second integrated mill at Anshan Works were also indefinitely delayed.

Chinese output in 1980 was some 37 million tons, and if the goal is 46 million tons by 1985, it means adding some 9 million tons, 3 million of which will come from the Baoshan plant and the rest from expanding existing facilities.

China announced in May 1979 that it would take at least two years to complete a review of the modernization timetable for the steel industry. This review probably will not be complete until well into 1982. Another interesting development in the Chinese industry is the closure of many small iron-making plants because they are considered uneconomical. Some nine of these were closed in the Gansu Province alone. The Chinese expected that their modernization and expansion plans for steel would be costly. However, when actual estimates were presented, they sent shock waves through the government. One compilation indicated that the total cost would reach almost $40 billion, with $17 billion for new plants, $12 billion for the expansion of existing facilities, and $10 billion for the development of iron ore.

The Chinese most probably will reach a production well over 40 million tons, and possibly 45 million, by 1985. However, it will be some time, at least to 1990, before the 60-million-ton figure is realized. In the interim, to meet the needs of the economy, steel imports will be increased, as will imports of iron ore. China will also remain a strong potential market for steel-industry plant and equipment. However, the great rush that took place between 1977 and 1979 has slowed appreciably. This reduction in the program will affect the world steel industry, since there will be at least 15 million tons fewer than had been projected for 1985. A much more-calculated program has been adopted by the Chinese that should result in a slower but perhaps a better-planned steel industry.

It is virtually certain that China will not be a factor in the export of steel products for years to come.

China and Japan

During the past decade, China has grown greatly in significance as a market for Japanese steel. Over the ten-year period it constituted the second largest outlet for Japanese exports, being surpassed only by the United States. In 1970, Japanese exports to mainland China were approximately 1.6 million tons; in 1971, they moved to almost 2 million tons, declining slightly to 1.7 million in 1972. After this, they increased rapidly, reaching 5.6 milion tons in 1978. Table 7–2 gives Japanese exports to its three principal consumers—the United States, China, and the USSR. The USSR has become a significant consumer since 1974, when exports to that country passed 1 million tons for the first time.

With the change in the Chinese government and the desire to industrialize as rapidly as possible in the late 1970s, imports from Japan rose quickly. Further, the Chinese, as previously indicated, engaged most of the Japanese

Table 7-2
Japanese Steel Exports to Principal Consumers: 1970-1980
(millions of metric tons)

Year	United States	China	USSR
1970	5.9	1.6	0,2
1971	6.3	1.9	0.4
1972	6.3	1.7	0.3
1973	5.3	2.7	0.6
1974	6.5	2.9	1.3
1975	5.7	2.8	1.1
1976	7.4	3.5	3.0
1977	7.6	4.5	1.3
1978	6.2	5.6	1.4
1979	6.2	4.5	1.7
1980	5.2	3.2	1.7

Source: Japan Iron and Steel Federation, *Monthly Report of Iron and Steel Statistics,* for the years involved.

steel companies to participate in the planned expansion that was to double output between 1978 and 1985. The reality of the situation forced a retrenchment in 1979 that was accompanied by a drop in steel imports from Japan; 1980 witnessed a further decline to 60 percent of the 1978 figure, and in 1981, imports fell to 2.2 million tons, a still further cutback. This has been a disappointment to the Japanese, but the three-year decline in activity in China is looked upon as a temporary phenomenon that will be reversed by the mid-1980s.

Since China did not build the steel industry that had been projected, it will continue to be a large market for Japanese exports; in fact, the longer the construction of its steel industry is delayed, the greater the market it will provide for Japanese steel. This is encouraging for Japan, since it is finding resistance to and competition with its export trade in many areas throughout the world. When world-steel activity is slack, the industrialized countries protect their home markets against the Japanese by limiting imports, and the Third World countries, which have been a very large market for Japanese steel, are installing much of their own capacity, thus limiting the amount of Japanese imports not only at the present time but also for well into the future. Thus, China will remain a very important outlet for Japanese steel since its home production of 38 million to 40 million tons for the next two or three years is woefully inadequate and must be supplemented from abroad. Here, again, the Japanese have a great advantage.

If one thinks in terms of the long run, once the Chinese steel industry is built up—and this will probably be after 1990—China will undoubtedly cut back on Japanese imports, and it could become a severe competitor to the Japanese in the Asian area. This is reminiscent of the European and U.S. relationship created by the Marshall Plan, in which the United States helped to build up the European industry and thus create its own competition. For Japan, this long-range consideration is overshadowed by the business that will be obtained in selling steel mills to the Chinese in the next decade.

Brazil

Brazil is second to China as a steel producer in the Third World, a position it has held since 1973 when it surpassed India. Its growth in steel production has been steady and rapid during the past decade, as it increased from 5.4 million tons in 1970 to 15.3 million in 1980.[10] In 1981, production had its first drop in a decade as it fell to 13.2 million tons.

In addition to its rank as a steel producer, Brazil has achieved a dominant position as a supplier of iron ore to the world. During the past few years, it has vied with Australia for number-one position as an exporter of iron ore, with both countries exporting close to 80 million tons of high-grade ore in 1980.

According to present plans, the Brazilian steel industry hopes to increase its raw-steel-production capacity to 22 million tons by 1985. Some of the more-optimistic steel producers place the figure at 25 million, but 22 million, or at most 23 million, tons seems to be a more-realistic achievement. The expansion plans for the industry have been on the drawing boards since the mid-1970s. In fact, at the 1975 meeting of the Brazilian Iron and Steel Institute, participants stated that steel capacity would reach 25 million tons by 1980. This goal proved to be too ambitious; however, production did double between 1974 and 1980, a significant achievement indeed.

The industry currently is dominated by three major integrated plants that in 1980 produced 8.7 million tons of the 15.3-million total, or 57 percent. These three plants—CSN, Cosipa, and Usiminas—each of which represents a separate company, are government owned as are a number of other smaller units. In 1979 and 1980, government-owned plants produced approximately 65 percent of total steel output, with the private sector providing the remaining 35 percent. Two of the largest private companies—Belgo-Mineira with a production of 874,000 tons in 1980 and Mannesmann with a production of 723,000 tons in the same year—are either owned or heavily participated in by European interests. ARBED is involved in the Belgo-Mineira property, and Mannesmann of Germany owns the Mannesmann facilities.[11]

The present expansion plans call for the construction of two fully integrated plants as well as the expansion of a number of existing facilities. The two new plants are Tuberao and Acominas. The former, to be located on the seacoast at a deep-water port, is currently under construction and should be in operation in early 1983. It will produce 3 million tons of steel to be converted into semifinished slabs. The company is a joint venture involving Siderbras with 51 percent, Kawasaki of Japan with 24.5 percent, and Italsider of Italy with 24.5 percent. The production is to be divided equally among the three partners. A second phase of the Tuberao plant will include additional steelmaking capacity and a hot-strip mill. However, this will not be operative until the late 1980s or possibly 1990.

The second integrated steel plant, Acominas, is partly complete and, when finished and put in operation in 1984, will produce 2 million tons of steel to be converted into long products. The plant is government owned and thus part of the Siderbras complex. Ownership of the plant is split up among the government-owned Brazilian steel company with 40 percent; the state of Minas Geras with 20 percent; CVRD, the iron-ore company, with 20 percent; and 20 percent foreign investment. Since CVRD is government owned, the public ownership of Acominas amounts to 80 percent.

A substantial expansion is planned for CSN at Volta Redonda. The plant produced some 2.5 million tons in 1980 that will be increased to 3 million tons in 1982 and further to 4 million tons by 1984. A number of smaller companies also have expansion plans, but they do not represent a very large aggregate tonnage.

Brazil is unique in terms of pig-iron output since it uses charcoal to produce a very large tonnage of iron. In 1980, this amounted to 4.7 million tons out of a total pig-iron production of 12.7 million tons, or 37 percent.[12]

Steelmaking technology includes the open-hearth process that contributed 9 percent of the 1980 production, the basic-oxygen process that accounted for 65 percent, and the electric furnace that produced 26 percent. A small fraction of 1 percent came form the Bessemer process. The new plants will use the oxygen process, which will increase its percentage of the total.[13]

Continuous casting has risen rapidly in Brazil in the last five years. In 1975, it accounted for 5.7 percent of steel production, or a mere 500,000 tons. By 1980, however, it had risen to 33.6 percent or 5 million tons.[14]

During the past ten years, Brazil's steel exports and imports have undergone a significant change as a result of its increased steelmaking capacity. Table 7–3 gives imports and exports, as well as the dollar value of each, for the years from 1974 to 1980.

The shift in terms of tonnage has been dramatic. From a ratio of 18 to 1 in favor of imports in 1974, it changed to more than 2 to 1 in favor of exports in 1980. In terms of dollars, it was about 20 to 1 in favor of imports in 1974, and by 1980, it had changed to a favorable export figure by about $60 million.

Table 7-3
Brazilian Imports and Exports of Steel and their U.S. Dollar Value:
1974-1980

Year	Imports		Exports	
	Tonnage (thousands)	Value (U.S.$)	Tonnage (thousands)	Value (U.S.$)
1974	4,197	1,417	236	69
1975	4,844	1,142	149	52
1976	1,059	498	262	65
1977	927	495	364	81
1978	733	415	935	210
1979	595	406	1,483	450
1980	657	492	1,498	551

Source: Instituto Brasileiro de Siderurgica, *Anuario Estatistico da Industria Siderurgica Brasileira, 1981* (Rio de Janeiro 1981).

Brazil is anxious to export steel. Its shipments to the United States increased tenfold within a five-year period from 1975 through 1980, rising from 37,000 to 410,000 tons. With an increase in steel capacity of about 6 million tons between now and 1985, exports will continue to rise as Brazil seeks to strengthen its trade position and to acquire hard currencies.

In terms of raw materials, Brazil has a tremendous supply of high-grade iron ore with reserves placed at more than 50 billion tons. However, it is necessary for the country to import a significant amount of coking coal. In 1980, 6.3 million tons of coal were processed into coke, of which 4.3 million tons were imported.

In the future, the Brazilian steel industry will continue to grow, although at a pace somewhat slower than had been hoped for.

India

India is the third largest producer among the countries of the Third World, preceded by China and Brazil. Unlike other countries in the Third World, India's steel output remained quite stable for the years from 1965 through 1974. In the former year, raw-steel production was 6 million tons, and it reached a high point of 7.1 million tons in 1974, the boom year for the world steel industry. In 1977, with the completion of the Bokaro Plant, India's steel output moved to 10 million tons. It remained at that level through 1979, dipping slightly to 9.4 million tons in 1980.[15] In 1981, it rose to 10.3 million tons.

The industry consists of two principal integrated steel complexes: the Tata Works, which is privately owned, that has a capacity of approximately 2 million tons and a government-owned complex, formerly known as the Hindustan Steel Company, that had three integrated steel mills at Bhilai, Rourkela, and Durgapur. The Indian Iron and Steel Company, which operated an integrated plant at Burnpur, with a capacity of 1.5 million tons, was privately owned until July 1972 when the government took over the management.

In 1977, the Hindustan Steel Company was reorganized into the Steel Authority of India Limited. It includes the three plants of Hindustan Steel and the Bokaro Steel Works and has combined a raw-steel capacity in excess of 7.0 million tons.

The total capacity for the Indian iron and steel industry as of 1981 was 11.4 million tons. Plans are under way to increase this amount to 14.0 million tons by mid-1983 with the construction of two new blast furnaces. Further long-range expansion plans will not come to fruition until the late 1980s or 1990. These include two additional steel plants to produce long products that, when completed, will each have 3 million tons of raw-steel capacity, bringing the country's steelmaking potential to 20 million tons by 1990. There is some talk about a third new plant for flat products with a raw-steel capacity of 3 million tons, but it is still in the discussion stage.[16]

India has abundant quantities of good iron ore; in fact, it exports significant tonnage. In 1978, exports amounted to 16.6 million tons of ore, averaging 61 percent iron content. In 1979, exports of this high-quality ore rose to 25.7 million tons out of a total production of 39.5 million tons. India, however, must import some 2 million tons of coking coal annually to mix with the domestic supply in order to form a satisfactory coke.

The country has a substantial trade in steel, with exports rising from well below 1 million tons in 1975 to 1.5 million tons in 1976. Imports fluctuated from a low of less than 500,000 tons in 1976 to a high of 1.5 million tons in 1979.[17]

India will have an additional 3 million tons of steelmaking capacity by 1985. However, growth after that, because of financial problems, will be slow. Between 1985 and 1990, 3 million tons or more may be added.

Most of India's steel, some 58 percent in 1980, was produced by the open-hearth method, with the oxygen and electric-furnace units contributing about 20 percent each. This will change somewhat, although not dramatically, in the years ahead with the installation of new capacity at Bokaro and Bhilai.

In addition to the integrated plants, a large number of small electric-furnace operations account for about 2.0 million tons of electric-furnace production. Further, a large number of rerolling plants must buy steel to be rolled down into a variety of products.

South Korea

Perhaps the most outstanding Third World country in terms of increased steel production in the past decade is South Korea. Output rose from 500,000 tons in 1970 to 10.7 million tons in 1981, and it could well increase by another 1 million to 2 million tons in the years immediately ahead.[18] The industry is dominated by the government-owned Pohang Steel Company that operates a fully integrated steel mill ranking with the most modern in the world.

Pohang's mill has recently been expanded to a capacity of 8.5 million to 9 million tons. It currently operates four modern blast furnaces, as well as one smaller furnace for the production of pig iron. Steelmaking facilities consist of two basic-oxygen shops, one with three 100-ton converters and a second with two 300-ton converters. Both continuous-casting and slabbing-mill facilities are in operation, as well as two hot-strip mills, one of which was brought on stream in 1981. There are also two plate mills, as well as a cold-reduction mill. Consequently, the output of the Pohang plant is principally in flat-rolled products although a new rod mill recently has been installed.[19]

The growth of this plant took place in several phases. The initital capacity was approximately 1.2 million tons. This was increased, subsequently, to 5.5 million tons, and in the last phase, completed in early 1981, capacity was expanded to between 8.5 million and 9 million tons. The Pohang plant, having been so recently constructed, is among the most modern in the world with a location on the seacoast, allowing it to take advantage of the deep water. It can receive large bulk-cargo carriers of at least 150,000 tons, thus keeping the cost of raw-material transportation at a reasonable figure. Like the Japanese, the South Koreans must import virtually all of their iron ore and coking coal. Thus, the seacoast location is extremely advantageous, if not necessary.

There is a plan to construct a second fully integrated steel mill on the south coast of South Korea. This will have an ultimate capacity of 6 million tons to be installed in two stages of 3 million tons each. As planned now, construction will probably start in 1984.[20]

In addition to Pohang Steel Company, the steel industry in South Korea has a number of privately owned electric-furnace plants, some of which have a significant production capacity. For example, Dongkuk Steel Mill Company Limited has a raw-steel capacity in excess of 600,000 tons, while Inchon Iron and Steel Company Limited has a capability approaching 400,000 tons. Another electric-furnace plant, Kisco, operates electric-arc furnaces capable of producing in excess of 200,000 tons.[21] As of the beginning of 1980, there were sixteen electric-furnace plants in the country with a total capacity of 2.6 million tons.

In 1981, of the 10.7 million tons of steel production, 80 percent was made by the oxygen process and 20 percent by electric furnaces. Continuous casting accounted for 32 percent of raw-steel production in 1980, but will increase in the years ahead as more planned units go into operation.[22]

South Korea has become an active exporter of steel in the last few years. In 1979 its exports amounted to 3.2 million tons, of which about 1.0 million went to the United States.[23] In the same year, it imported 2.7 million tons, of which 2.0 million came from Japan. The construction of the additional steelmaking capacity at Pohang, as well as the second hot-strip mill, has reduced South Korea's dependence on Japanese imports, since more than one-half of the tonnage from Japan was in the form of flat-rolled products. South Korea not only will be able to supply its own needs for this product, but also will become a significant exporter. In 1981, Pohang shipped 1.0 million tons to Japan as well as substantial amounts to other Far Eastern markets.

There is little doubt that South Korea will continue to export considerable steel in the next few years, for although projections concerning domestic needs are high in the late 1980s, before that time much of the additional capacity will find its way into the world market.

Mexico

The Mexican steel industry, along with several others in Latin America, has grown significantly in the last ten years. Raw-steel output in 1970 was 3.9 million tons, and it increased slowly but steadily until it reached 7.2 million tons in 1980.[24] In 1981, it rose to 7.4 million tons. The industry is composed of five principal producers and a number of small electric-furnace operators. Output in 1980 among the steel makers, in millions of tons, is shown in the following list.[25]

Altos Hornos de Mexico	2.3
Hylsa Group	1.6
Fundidora Monterrey SA	1.0
Sicartsa	0.8
Tamsa	0.4
Semi-integrated producers (some 40 mills)	1.1
Total	7.2

This production was somewhat short of actual capacity that is between 9 million and 10 million tons.

A number of plans for expansion will be carried out in the next five years. Altos Hornos, the largest company, expects to reach 4.2 million tons. Hylsa hopes to increase its capacity to 2.8 million tons by 1985 in an ambitious program that will witness significant increases in steel-making and finishing capacity at both the Monterrey and Puebla plants.[26]

Sicartsa, a new government owned steel-mill complex built on the west coast of Mexico in the late 1970s, will add more capacity. Transportation and other problems restricted its output in the first few years, however, in 1980, it reached 800,000 tons with a capacity of approximately 1.1 million tons. An expansion program is under way at this plant to add one million tons to double its capacity during the next two to three years. Tamsa, a producer of pipe, is installing some 250,000 tons of added capacity to increase its potential in the next two to three years to 750,000 tons.

The third largest producer, Fundidora in Monterrey with a capacity of 1.5 million tons, has no plans to expand in the next few years.

The Mexican industry by 1985 should be capable of producing approximately 13 million tons. This amount of steel will be consumed domestically since a high rate of economic growth has occurred in Mexico, due principally to the discovery of large oil reserves. This inspired economic forecasters to project at least a 7 percent annual growth, and the steel industry has responded by planning increased capacity.

Mexico, unlike most of the Third World countries that have built up a steel industry, exports very little of its steel output; in fact, it is presently a significant net importer of steel. In 1980, production of raw steel, as indicated, was 7.2 million tons, while consumption in terms of raw-steel equivalent was over 10 million tons. Thus, in raw-steel equivalent, over 30 percent of the steel used in Mexico had to be imported.

In terms of raw materials, on the one hand Mexico is virtually self-sufficient in respect to iron ore, with reserves that will last for another twelve to fifteen years. On the other hand, coking coal has to be imported. This comes almost exclusively from the United States, which provided 548,000 tons of coking coal in 1980 and some 510,000 tons in 1979.[27]

Only three of Mexico's five major companies have blast-furnace operations—Altos Hornos, Fundidora, and Sicartsa. In the expansion program, no new blast furnaces will be constructed. The program at Hylsa will be based on the use of DRI which has been Hylsa's raw material since 1957, when it established the first direct-reduction plant in the world to function on a continual-production basis. Sicartsa will use direct-reduced material from the HYL process for electric furnaces in the expansion plan, as will Tamsa.

In 1975, a relatively small amount of Mexican steel—namely, 700,000 tons, or 13.2 percent—was continuously cast. This increased to approximately 2.1 million tons in 1980, or almost 30 percent.[28] The expansion program for the years ahead will employ continuous casting so that by 1985, continuous casting should increase significantly.

The major portion of Mexican steel is produced by the electric furnace. In 1980, it accounted for 42 percent of the output. Basic oxygen was responsible for 37 percent, while the open hearth accounted for 21 percent.[29] In the years ahead, the electric furnace will produce an even higher percentage, since the expansions at Sicartsa, Hylsa, and Tamsa are to be based on the electric furnace. Mexico was the pioneer in the production of DRI with the HYL process, which dates back to 1957. The electric-furnace expansion will be based on DRI produced by the HYL process from its new development called HYL III. By 1985, more than one-third of the steel produced in Mexico will be based on DRI.

The Mexican steel industry, in terms of capacity, is 65 percent government owned, with three companies—Altos Hornos, Fundidora, and Sicartsa—combined under the Sidermex Holding Company. Hylsa and Tamsa, as well as the small electric-furnace operators, are in the private sector.

In terms of trade, with few exceptions, imports must be licensed, and the steel industry recommends the licensing and does the importing, so that steel coming into the country is sold through Mexican steel companies. Thus, the industry has a high degree of government protection for its market.

This is one of the reasons plans for expansion have been made and can be carried out with the reasonable assurance that the steel produced will be sold domestically at profitable price levels.

Given the needs of its economy, the country, when it solves its immediate economic problems, will absorb all of the steel it can produce in the next few years. It will continue to be a net importer with very little in the way of exports. Evidence of this is the drop in exports to the United States in recent years. These fell from a high of 224,000 tons in 1977 to 67,000 tons in 1980.

Taiwan

The steel industry in Taiwan has grown rapidly during the 1970s as production rose from 470,000 tons in 1971 to 4.2 million in 1980. Growth was gradual between 1971 and 1977, as it rose to 1.8 million tons in the latter year. In 1978, with the initial operation of China Steel's major project, output rose to 3.4 million tons. Since that time, it has increased significantly to the 4.2 million-ton figure.[30]

China Steel is a new, fully integrated plant whose first phase involved a steel-production capacity of 1.5 million tons. The plant has coke ovens, a 4000-ton-a-day blast furnace, two basic-oxygen converters, continuous-casting machines, a plate mill, and a rod and bar mill. The second stage of the plant's growth, to be completed in 1982, will push capacity to 3.2 million tons. It consists of a 5,000-ton-a-day blast furnace, enough coke ovens to supply it, a third basic-oxygen vessel, and in the finishing end, a 68-inch hot-strip mill, as well as a cold-reduction mill of the same proportions.[31] China Steel not only will provide for the sheet and strip needs of Taiwan but also will have excess capacity to ship steel abroad to the Asian market.

The country is devoid of the essential raw materials for steel production so it must import all of its ore and coking coal. The ore comes principally from Australia with smaller amounts from South America, while the coal is imported from Australia with smaller amounts from the United States and Canada.

In addition to China Steel, a number of electric-furnace operators throughout the country have an aggregate capacity of between 2.5 million and 3 million tons. The feed for this segment of the industry is scrap, much of which is imported from the United States, and a significant amount is developed through shipbreaking operations in Taiwan. Imports from the United States increased dramatically in the years from 1978 to 1980. They rose from 304,000 tons in 1978 to 634,000 in 1979 to 990,000 in 1980.[32]

The proportions of steel continuously cast and produced by the basic-oxygen and electric-furnace processes will change significantly in 1982 with the completion of China Steel's second stage of development. In 1980, two-thirds of the steel was produced by the electric-furnace process and one-third by the oxygen method. This will be considerably altered when China Steel adds another 1.7 million tons of basic-oxygen capacity during 1982. In terms of continuous casting, 1.8 million tons, or 44 percent, of the total steel produced in 1980 was continuously cast.[33] This will unquestionably increase as a result of the aforementioned expansion.

Total capacity for the nation will be about 6.0 million tons by the end of 1982, when China Steel's current expansion program is completed. There will be a further increase in capacity during the late 1980s when and if China Steel adds an additional 3 million tons to its productive capability. The company had hoped to accomplish this by 1985, but at present, it apparently will be delayed for at least two to three years.

Taiwan has been involved in steel trade to a significant extent, when one considers the size of its production. In 1970, imports were 650,000 tons; this increased substantially during the decade, reaching 2.1 million tons in 1979. The country exported 345,000 tons in 1970. This figure fluctuated somewhat through 1975. However, by 1979, it had risen to 1.5 million tons, causing concern among Japanese steelmakers.[34]

Taiwan will continue to export steel and most probably will increase the tonnage, principally in sheets—as its new facilities are brought into production in late 1982.

Argentina

The steel industry in Argentina produced 2.7 million tons of raw steel in 1980, which was 16 percent below the all-time high of 3.2 million tons in 1979.[35] Judging from the number and size of the iron- and steelmaking facilities now in place, the capacity of the Argentinian industry could well be as high as 6.5 million tons. However, the rate of utilization is low due to bottlenecks in production and the market in Argentina, which has been relatively weak in the past few years despite an increase in demand throughout most of Latin America. Compared to the output of Brazil, which doubled between 1974 and 1980, Argentinian output increased at a very slow pace, moving from 2.4 million tons in 1974 to 2.7 million tons in 1980.[36]

The two dominant companies in the country are Somisa and Acindar. The largest is Somisa, a government-owned entity that has a potential, if the bottlenecks were broken, to produce 2.5 million tons annually and possibly a bit more. The company has four coke-oven batteries that can produce 1.3 million tons and two blast furnaces, one with a 2,300-ton-per-day capacity and a second with a 3,600-ton-per-day capacity. The steel-making department has three 200-ton basic-oxygen converters that have a capacity of about 3 million tons a year. In addition, five open hearths, with 235-ton capacity each, have an annual total capacity of 1.1 million tons. Strictly speaking, in terms of furnace size, Somisa should be able to produce 4 million to 4.2 million tons; however, it has never approached even half of this amount. The company produced 1.5 million tons in a depressed market during 1980, but even in the record output year for the country, 1979, Somisa's production was still not 2 million tons. The company operates continuous-casting facilities for billets and blooms that are offered for sale; however, the market for much of this product was Acindar, which is now producing its own steel.

Somisa is confined primarily to light, flat products from its 68-inch hot-strip mill, currently producing 1.4 million tons from 1.6 million tons of slabs. The slabbing facility constitutes a bottleneck since it can only produce about 1 million tons, forcing Somisa to import slabs to make up the difference. In addition to hot- and cold-rolled sheets, tinplate is also produced from cold-reduced material obtained from Propulsura, a company that operates a cold-reduction mill based on imported hot-rolled coils.

Somisa is studying several plans to expand its capacity to 4.3 million tons. One of these is the installation of the KMS process, developed by Klockner, that will allow the use of more scrap or direct-reduced material in the basic-oxygen converters and, thus, increase steel output. Another possibility, although somewhat remote, is a third blast furnace. If KMS is chosen, it will probably require direct-reduction facilities.[37] The company has bought a continuous caster to expand its output of slabs and has on the premises a plate mill and a slabbing mill that have yet to be installed. If more iron units can be produced, there should be no problem in achieving the 4.3 million tons. However, at the present time, the company could not finish this amount of steel and would need a new hot-strip mill to do so. The installation of such a facility would be costly.

Acindar is the second largest steel producer in Argentina and devotes its output to long products. Recently, it merged with Santa Rosa and Gurmendi. Acindar has an electric-furnace capability of 900,000 tons per year. This comes from two of its three furnaces; one unit is idle because there is neither enough iron to feed it nor enough continuous-casting capacity to process the steel. If both these shortcomings are remedied, output could readily be 1.3 million tons a year. The iron units can be obtained by increasing the gas-reformer capacity on the Midrex direct-reduction plant, which would bring output of that unit to 650,000 tons. This, coupled with another caster, will produce some 300,000 tons more of billets by 1985. Santa Rosa has 250,000 tons of electric-furnace capacity and produces bars. The total Acindar complex can produce 1.15 million tons of steel and, with relatively minor improvements, can expand this to 1.5 million tons by 1985.[38]

There are a number of smaller steel operations including Dalmine, a producer of seamless tubes, with 400,000 tons of electric-furnace capacity and an annual output of some 300,000 tons of seamless tubes. Another mill, Altos Hornos Zapla, is an installation operated by the military with five charcoal blast furnaces and a Thomas converter with a capacity of 250,000 to 300,000 tons of steel. It also has an electric-furnace capability to produce between 50,000 and 60,000 tons. Products from this unit include shapes, bars, wire rods, and some specialty steel.

A new plant, Sinder Sur, is under construction. This will be devoted to the production of DRI with the installation of an HYL III unit. The plant is located in the south where gas is available and where a harbor is capable of bringing in big ships with large quantities of ore. Other plants have been proposed, such as Sidinsa, which was to have 1.7 million tons of steel and a hot-strip mill. Recently, the hot-strip mill, if it is constructed, has been assigned to Somisa, so apparently the Sidinsa plant has been postponed indefinitely.

In 1975, the steel industry in Argentina employed continuous casting for 26 percent of its output that, at the time, was equal to 600,000 tons. By 1980, 1.4 million tons were continuously cast, or some 53 percent of total production. This will increase in the years ahead, depending on the growth of the entire industry. Somisa plans to install more continuous casting as does Acindar. In 1980, the 2.7 million tons of Argentinian steel production represented a 54 percent contribution by the electric furnace, 26 percent by the oxygen converter, and 17 percent by the open hearth. Argentina is one of the few countries still employing the Thomas process, which contributed about 3 percent.[39] In the years ahead, the open hearth will dwindle and will be replaced by the oxygen process and electric furnace.

In regard to raw materials, almost all of the coal is imported, coming from the United States and Poland, while a significant amount of iron ore is brought in from Brazil.

In terms of trade, Argentina has been a net importer of steel for a decade. Imports reached their high point in 1975 with over 2 million tons. Since then they have tapered off considerably, falling well below the million-ton mark in subsequent years. In recent years, a significant portion of the imports has consisted of steel for further processing, such as slabs and hot-rolled coils to be cold reduced. Exports have been low for many years, surpassing the half-million-ton mark in 1973, after which they declined. For the next few years, Argentina is unlikely to be a major exporting country even by Third World standards. From time to time, there may be a surge such as the one in 1978 when exports to the United States totaled 239,000 tons, almost half of which were in hot- and cold-rolled sheets. Two years later, they dropped to a low of 18,000 tons.

Venezuela

In the late 1950s, the Venezuelan government organized a steel company known as Sidor, which was to build an integrated steel plant capable of producing more than 1 million tons annually. Since the country has abundant iron ore but virtually no coking coal, the ironmaking segment of the plant was based on Tysland-Hole electric-reduction furnaces that provided hot metal for four 300-ton basic open-hearth furnaces with a total rated capacity of 1.2 million tons. The plant operated at a high rate of capacity during the 1970s, as annual production averaged a little less than 1 million tons, with the exception of 1977 and 1978 when it dipped below 800,000 tons.

In 1975, expansion plans were drawn up involving the installation of several direct-reduction units to produce DRI for two large electric-furnace shops, one with six 200-ton furnaces and the other with four 150-ton furnaces, which would make Sidor the largest electric-furnace installation in the world. Completed in 1979, the full capacity of the new plant was between 3.6 million and 3.8 million tons, raising the total capacity of Sidor to almost 5 million tons. The output for 1980 was approximately 1.6 million tons, or less than 50 percent of capacity. In 1981, output increased to 1.8 million tons.

The expanded plant is based on DRI produced from four different processes—Midrex, HYL, FIOR, and United States Steel's HIB process. Start-up problems have plagued the company to a great extent. However, many of these apparently are on the way to a solution so that, by 1983, it is hoped that output of raw steel will be 3 million tons and will increase to 4 million tons by 1984. Unlike other countries where expansion is in the planning or construction stage, Venezuela has steelmaking capacity in place, and as production practices improve, it is reasonable to expect that Venezuelan steel output will increase substantially.

The production of DRI is made possible in Venezuela by the abundant supply of cheap natural gas that is needed to operate the direct-reduction units. These units have a total rated capacity of approximately 5 million tons. However, total utilization has never been achieved.

In addition to Sidor, a small electric-furnace producer called Sivensa operates two plants, one at Caracas and the other in Barquisimeto. Total capacity is approximately 200,000 tons.

During the past two to three years, considerable discussion has concerned the erection of a new steel plant in the western part of Venezuela near Maracaibo. It is known as the Zulia Steel Project, or Corporzulia. As first projected, it was to include a blast furnace with a 1-million-ton capacity and a basic-oxygen converter with 1.2 million tons of steel capacity, continuous casting, and rolling mills for bars, wire rods, and structural sections. The plan has since been considerably modified and, at present, is reduced to a structural mill capable of producing 250,000 tons to 300,000 tons of medium-size sections. There is also talk of erecting a seamless-pipe mill with approximately the same capacity. The plant in Maracaibo will be state owned, and in the years ahead it may be fully integrated.

The construction of a plant in western Venezuela at this time is based on political rather than economic motives. Sidor is operating at less than 50 percent of capacity and incurring large losses, some $400 million in 1981. Economically speaking, the Sidor plant should be brought to its full potential before another facility is constructed. Further, it would be far more practical to put a structural mill and seamless-pipe mill at Sidor where there is ample steel and space to accommodate these units. However, the political realities in Venezuela seem to dictate otherwise.

At the present time, Venezuela, in spite of its increased steel capacity, imports more than 1 million tons of steel products. These are controlled by the government with the advice of Sidor. As yet, Venezuela is not a significant factor in the export of steel. However, if Sidor reaches its potential, Venezuela will have large quantities of steel products to export, since the country's economy cannot absorb the full potential of Sidor.

Total steel production in Venezuela is provided two-thirds by modern electric furnaces and one-third by the open-hearth process. No oxygen steel-making capacity exists in the country. More than half of the country's output is processed by continuous casting, and as production increases, the percentage of continuously cast material will grow.

Venezuela, having exported as much as 26.3 million tons of iron ore in 1974, is a large supplier of that material to the world steel industry; of this, 15.5 million was taken by the United States. Since that time, sales to the United States have dropped to 3.5 million tons in 1980, and the total to the world has declined to 11.9 million tons.[40] Venezuela is very much interested in developing markets for its iron ore. However, in competition with Brazil, it has a distinct disadvantage since its harbor depths restrict ship size to 50,000 and 60,000 tons as compared with 250,000-ton carriers from Brazil. The difference in freight charges to Europe is approximately $4 a ton.

Sidor, heavily burdened with debt, has sustained losses over a number of years and has had to depend on the government to compensate for them. It is hoped that by 1985, with a revised debt structure and an improvement in production, the losses may be turned into profits.

Other Third World Countries

The foregoing analysis of the eight largest steel producers in the Third World is not meant to be exhaustive but more of an indication of the present state of the steel industry in those countries as well as of their plans for the future. Many other countries in the Third World are engaged in the production of steel; however, their output is much smaller than that of the eight countries discussed. The one exception is North Korea. Neither accurate statistics on North Korea's production nor much information on its future plans is available. The International Iron and Steel Institute has estimated production for North Korea for more than fifteen years. In 1965, the figure was placed at 1.2 million tons and graduated steadily upward since that time with virtually no interruption except in 1975. By 1980, the estimate had reached 5.8 million tons.[41] This presumed a growth in output that corresponded, to some extent, to the growth in Third World steel production.

A number of other countries in the Third World plan to expand their steelmaking capacity during the next few years. They include Chile that hopes by 1985 to expand its current output, which was 743,000 tons in 1980, to approximately 1 million tons.[42] The program is a sound one because it involves the replacement of some outmoded facilities, as well as bringing the entire plant into balance. Currently, funds are in short supply and the work has been held up.

Nigeria, which heretofore had very limited steel production, has just completed a 1.1-million-ton plant based on DRI, electric furnaces, and continuous casting. The steel will be processed into a variety of bar products. A second plant is now under construction that will employ conventional steelmaking practices including a blast furnace and oxygen steel making with an ultimate capacity of 1.5 million tons. Discussions have begun on a third plant based on DRI and electric furnaces that, if brought to fruition, will produce 2 million tons of flat products. However, this plant would not be complete until well into the late 1980s. Nigeria will have two plants operating by 1985 with a capacity of approximately 2.6 million tons.[43]

Egypt has had production problems with its principal plant that was originally designed for 1.5 million tons. The management has secured technical assistance and hopes to improve this plant's operation in the next few years and raise it to 1.1 million tons. There is also a plan for a new plant, based on DRI, with an 800,000-ton steelmaking capacity to be turned into bars.

Libya has contracted for two direct-reduction units with an aggregate capacity of 1.1 million tons. A steel plant is also under contract to produce a total of 1 million tons in two melt shops. The ultimate products will be plates, bars, and structural sections.

Algeria, like Egypt, has had production problems and recently secured technical assistance that may increase the output of its principal plant to 2.0 million tons.

Saudi Arabia is currently completing the construction of its direct-reduction electric-arc furnace complex, which will produce 800,000 tons of raw steel.

Indonesia currently has 500,000 tons of electric-furnace capacity fed by DRI. Within the next two years, if present plans hold, electric-furnace capacity will be increased to 2 million tons to be processed through a continuous slab caster and a hot-strip mill.

The proposed plant in the Philippines as well as that in Malaysia have already been mentioned. There are a number of other projects that will be undertaken in Third World countries, such as Bangladesh and Pakistan, during the next few years. Many of them are relatively small and will not be in operation until 1985.

Future Growth in the Third World

The Third World is not a group of homogeneous countries; on the contrary, there is a wide divergence of economic development and financial resources among the nations that constitute the Third World. The oil-producing countries, members of OPEC, are all Third World countries. These nations have the greatest financial resources, while other countries are poor and in the rudimentary stages of economic development. Further, a country like Brazil has a greater degree of economic development than some countries that are not officially considered part of the Third World.

The diversity of economic status will have a direct bearing on the future development of steel production and capacity in the various countries. As indicated, nine countries have steel outputs ranging from 2 million to 37 million tons, while all the others have productions ranging from less than 100,000 to approximately 1.5 million tons.

Much publicity has been given to the growth of the steel industry in the Third World, and optimistic projections have been made, claiming that the future of the world steel industry lies among these countries. As previously indicated, there has been a substantial percentage growth in steel output in the Third World countries, particularly between 1974 and 1980, as output increased by 67 percent. Based on this growth, studies have attempted to prove that the Third World has a cost advantage in steel production over the industrialized countries and, so, should gain a greater share of future steel output as the industrialized countries lose it. Unfortunately, in some of these studies, the conclusion is presupposed, and the inquiry is an attempt to justify it.

The question of Third World growth in steel production has been examined on a country-by-country basis to ascertain precisely whence it will come in the next decade. The greatest potential for growth is in those countries that currently produce significant tonnages of steel, although some of the oil-rich Middle Eastern countries will become producers of steel either by installing steelmaking capacity for the first time or by adding to that small amount they already have. To speak generally in terms of Third World growth is not particularly satisfactory or enlightening, since obviously any increase will come from the individual countries. An analysis based on individual countries indicates the growth potential of the Third World by 1985. Total capacity to be added by that year in the eight largest countries could amount to 21 million tons. It should be pointed out that some inactive capacity in countries such as Venezuela and Argentina will be brought into production but that this will not add any new capacity. The remainder of the Third World, particularly Middle Eastern countries such as Libya, Saudi Arabia, Egypt, Iran, Iraq, and Nigeria, will install steelmaking capacity, most of which will be based on DRI. As indicated, the units under construction, on order, or actively under consideration are relatively small, with the largest in Nigeria and Libya. For the entire Middle East and Nigeria, the maximum increase in steelmaking capacity by 1985 will be somewhere between 4.5 million and 6 million tons. This will bring the total for the Third World, exclusive of some Far East countries such as Indonesia and Pakistan, to 26 million to 27 million tons. Other areas in South America such as Chile, Peru, and Colombia will add small amounts. However, in almost all of these countries, financial difficulties will restrict the total and slow down the timing.

In the entire Third World, a generous estimate of all new capacity by 1985 would be about 29 million tons. If past performance is taken as a criteria in terms of plant installations, the figure could well be reduced to 25 million tons. Should all of the proposed tonnage be put into operation, it could represent a production increase of some 30 percent over the 1980 figure, or a total of 130 million tons. Given a world production of 770 million to 800 million tons by 1985, this would amount to approximately 15 percent. Thus, the growth of the Third World's steel industry in the next few years will be small in relation to total world production. In the period beyond that, stretching to the year 2000, there unquestionably will be more steel capacity installed in the Third World. However, projections as to its amount are risky.

In assessing the competitive impact on steel producers in the industrialized world, one must consider that (1) Third World capacity has been increasing at a slower rate than originally anticipated; (2) most of the new capacity is intended to satisfy home-market requirements, displacing imports largely obtained from producers in the industrialized world; and (3) while some of the export tonnage generated by new Third World capacity will find its way into industrialized-country markets, some will be traded within the Third World, displacing steel now obtained from industrialized countries.

Accordingly, the overall competitive impact of Third World steel growth through the current decade will likely be less than some observers expected, based on the optimistic projections originally formulated and the publicity that Third World steel activity has been given.

Many of the countries throughout the Third World assert that they need steel production at home. For example, South Korea projects a need for more than 25 million tons in the late 1980s. Similar statements have been made by the Brazilians. However, it is a matter of record that a portion of the new capacity, installed in these countries in the last few years, has provided tonnage for the export market.

In spite of the fact that most of the steel produced in the Third World will be used domestically, particularly as demand in those countries grows, a significant amount will continue to be shipped to international markets by the larger producers. The Third World will participate in international steel trade, and it is most probable that some countries such as Brazil, South Korea, and Taiwan will establish strong positions in this area.

Cost Competitiveness

To establish a position in international trade, the industry of a country must be cost competitive with other major steel-producing companies. The alternative is a subsidy to allow the company to compete in terms of price, even though its costs are out of line. Some Third World countries are cost competitive with the industrialized world due to extremely low labor rates and the recent installation of new, government-financed equipment. Two outstanding examples of the cost competitiveness are South Korea and Taiwan where labor costs are less that $4 an hour and where new, fully modern plants have been built in the past five years. South Korea and Taiwan must import their raw materials; nevertheless, their overall costs are low enough to make their steel competitive in the international market.

Other countries are price competitive primarily because their steel industry is subsidized by the government and not because of any inherent efficiencies. A number of these government-owned companies are losing vast sums of money, like Sidor in Venezuela that posted a $400 million loss in 1981. Many others are in the same circumstances and need these subsidies indefinitely if they are to survive. Very often, the plan is to subsidize the company for a limited number of years and then to allow it to operate on its own. However, the losses, and in most cases these are due to huge interest charges, have been so great that, practically speaking, subsidies must be continued if the company is to survive, and in most instances, government policy strives to keep the company in operation in order to maintain employment. Further, as previously indicated, many of the plants are built because of political rather than economic considerations, and it is doubtful if they will ever be profitable except for years when steel demand on a worldwide basis is exceptionally high.

Raw Materials

Many of the Third World countries that are developing steel industries depend on imports for one or the other of the basic raw materials required for the production of steel. For example, South Korea and Taiwan must import both iron ore and coal. Mexico is dependent on the outside for coal as is Brazil and Argentina. Countries that install direct-reduction units and electric furnaces usually do so because they have adequate supplies of gas to operate those units. This is true in Venezuela, Nigeria, Saudi Arabia, and Libya. Some of these countries, however, must import the iron ore.

In general, very few Third World countries except India have all the raw materials necessary for steel production so that they will continue to look to the outside, into the indefinite future, for whatever they lack.

Technology

The steel companies in the Third World, which have grown in the past five years and will continue to do so in the next five, are employing modern technology in most instances. In the production of steel, both the coke-oven-blast-furnace-basic-oxygen complex and the DRI-electric-furnace complex have been adopted. Blast furnaces like those installed in South Korea and Taiwan are large and efficient. Electric furnaces with DRI units providing the feed are of the most modern design, as witnessed in Mexico, Nigeria, and Venezuela. Continuous casting has been adopted by most plants.

In terms of rolling mills, problems are developing. The very sophisticated, electronically controlled units that have been installed in the industrialized world have presented difficulties in many of the Third World countries due to the lack of skilled personnel to operate and maintain the units. A number of Third World steel producers have suggested that they install less-sophisticated rolling and finishing equipment. As one steel-company official put it: It is better for us to have a rod mill operating at 6,000–7,000 feet a minute that can be maintained than one operating at 15,000 feet a minute that is too complicated for our personnel to operate and maintain.

Ownership

By far, the majority of steel plants, in operation or planned, in the Third World are government-owned. There are a few exceptions in countries like Brazil where approximately 35 percent of the industry is privately owned. Mexico also has a thriving private sector in the steel industry as does Argentina. In a number of instances, there are joint ventures with outsiders like in Brazil, where the Tuberao plant now under construction is owned jointly by the Brazilian government, the Italians, and Kawasaki Steel Company in Japan.

In a number of smaller plants in countries such as Saudi Arabia and Iraq, suppliers of equipment have taken an ownership position.

Notes

1. Proceedings of the Second General Conference of the United Nations Industrial Development Organization, Lima, Peru, March 1975.

2. Ibid.

3. IISI, *Steel Statistical Yearbook 1981,* p. 2.

4. Much of the information in this section on China, unless otherwise noted, was developed from internal memoranda circulated in the Japanese steel industry, as well as memoranda circulated among the management of Broken Hill Proprietary Co. Ltd., in Australia. The Japanese have been

heavily involved in the entire Chinese steel program and thus have more information concerning it than is available elsewhere.

5. "The Iron and Steel Industry of China," *Metal Bulletin* (1978):83.

6. Ibid.

7. IISI, *Steel Statistical Yearbook 1981,* p. 3.

8. Interviews with Japanese steel-industry executive personnel.

9. Ibid.

10. IISI, *Steel Statistical Yearbook 1981,* p. 2.

11. Instituto Brasileiro de Siderurgica, *Anuario Estatistico da Industria Siderurgia Brasileira 1981* (Rio de Janeiro, 1981), p. 20.

12. Ibid., pp. 32 and 33.

13. IISI, *Steel Statistical Yearbook 1981,* p. 5.

14. Ibid., p. 7.

15. Ibid., p. 3.

16. Interview with executive personnel of Steel Authority India Limited.

17. IISI, *Steel Statistical Yearbook 1981,* pp. 23 and 24.

18. International Iron and Steel Institute, *Monthly Statistical Reports.*

19. Interview with executive personnel of Pohang Steel Company.

20. Ibid.

21. *Iron and Steel Works of the World* (London: Metal Bulletin Books, Ltd.), pp. 513–519.

22. IISI, *Steel Statistical Yearbook 1981,* pp. 5 and 7.

23. American Iron and Steel Institute, *Annual Statistical Report for 1980,* p. 41.

24. IISI, *Steel Statistical Yearbook 1981,* p. 2.

25. Discussions with executive personnel of HYLSA, S.A.

26. Ibid.

27. *Coal International,* February 1980, Zinder-Neris, Inc. and *Coal Statistics International,* February 1981, McGraw-Hill.

28. IISI, *Steel Statistical Yearbook 1981* (Washington, D.C.), p. 7.

29. Ibid.

30. Ibid., p. 3.

31. Conversations with employees of United States Steel Corporation's Engineering and Construction Division.

32. "Iron and Steel Scrap, Monthly" for January 1979, 1980, and 1981. *Mineral Industry Surveys,* U.S. Department of the Interior, Bureau of Mines, Washington, D.C.

33. IISI, *Steel Statistical Yearbook 1981,* p. 7.

34. Ibid., pp. 23 and 24.

35. Ibid., p. 21.

36. Ibid.

37. Information on Somisa facilities and projected plans was obtained from interviews with executive personnel of Somisa.

38. Data on Acindar were obtained from interviews with executive personnel of that company.

39. IISI, *Steel Statistical Yearbook 1981,* pp. 5 and 7.

40. U.S. Department of the Interior, Bureau of Mines, *Minerals Yearbook,* vol. 3 (Washington, D.C., 1974).

41. IISI, *Steel Statistical Yearbook 1981,* p. 3.

42. Ibid.

43. Data on Nigeria, Egypt, Libya, Algeria, and Saudi Arabia were obtained from conversations with companies involved in the installation of steel-mill equipment in those countries. They include Midrex Corporation, United States Steel Corporation, and Voest-Alpine.

8

The Soviet Bloc Faces a Declining Growth Rate

The steel industry in the USSR and the Eastern European Communist countries moved counter to the world trend in the last half of the 1970s. As steel output fell sharply in the industrialized world in 1975, it increased significantly in the Soviet bloc, not only in 1975 but also through 1978, rising from a total of 185 million tons for the area in 1974 to 211 million tons in 1978. In both 1979 and 1980, it fell to 209 million tons.[1] The principal contributor to tonnage was the USSR, which became the largest steel producer in the world in the early 1970s, surpassing the United States for the first time. It produced 136.2 million tons in 1974 compared to 132.2 million for the United States, and since that time, has been the world leader in steel production.

The remainder of the Eastern bloc, including Bulgaria, Czechoslovakia, East Germany, Hungary, Poland, and Rumania, has also witnessed a continual growth in steel output from 1974 through 1980. Growth slowed down appreciably after 1978 in the Eastern European Communist countries as the total for the area, exclusive of the Soviet, rose but slightly from 59.9 million tons in 1978 to 60.4 million in 1979, and reached 61.2 million in 1980. In 1981, production for these countries fell to 58 million tons because of a drop in Poland's output.

The USSR experienced its first decline in more than twenty years in 1979 as output fell to 149.1 million tons from the 1978 high of 151.4 million tons. Production declined further in 1980 to 147.9 million tons.[2]

In tracing the growth of the Soviet output, there was a remarkably consistent increase on an annual basis of from 4 million to 6 million tons from 1966 through 1975, as table 8-1 indicates. As can been seen from table 8-1, the growth rate declined between 1975 and 1977, resuming its upward trend in 1978, after which production declined. Over the same period, the Eastern bloc of European Communist countries witnessed a steady annual growth.

Technology

In terms of technology, the USSR did not keep pace with the Western world, since the major portion of its steel was and still is produced by the

Table 8-1
Growth of Soviet Bloc Steel Production: 1966-1980
(millions of metric tons)

Year	USSR	Eastern Bloc
1966	96.9	30.5
1967	102.2	33.1
1968	106.5	34.9
1969	110.3	37.3
1970	115.9	40.1
1971	120.6	42.4
1972	125.6	45.0
1973	131.5	46.8
1974	136.2	48.9
1975	141.3	51.3
1976	144.8	54.2
1977	146.7	57.5
1978	151.4	59.9
1979	149.1	60.4
1980	147.9	61.2

Source: International Iron and Steel Institute, *Steel Statistical Yearbook 1981* (Brussels), p. 3.

open-hearth process. In 1966, 82 percent of the output was accounted for by the open hearth and only 9 percent was from the oxygen process. By 1974, the picture had changed somewhat as the open hearth accounted for 65 percent as opposed to 25 percent from the oxygen process and 10 percent from the electric furnace. In 1980, the open hearth was still the dominant producer with 60 percent as compared with 29 percent from the oxygen process and 10 percent from the electric furnace. This fell short of the plan that was to have 34 percent of the steel output from oxygen furnaces and 19 percent from electric furnaces by 1980.

The Eastern European Communist countries followed the same pattern. In 1974, with the exception of Bulgaria, the open-hearth process provided most of the steel for the Eastern European Communist bloc. By 1980, the proportions had not changed dramatically, as indicated in table 8-2.

In terms of continuous casting, in 1970, the Soviets had some 5 million tons, or 4.3 percent, of their output continuously cast, Czechoslovakia had 0.4 percent, and Poland had 2.2 percent. By 1975, the USSR had increased its continuous-casting capability to 6.9 percent, Czechoslovakia had increased to 0.5 percent, Poland remained at 2.2 percent, and Hungary and East Germany installed continuous-casting capacity to the extent of 21.1 percent

Table 8-2
Crude-Steel Production, by Process, for Eastern Bloc Countries:
1974 and 1980
(percentage of total)

Country	Open Hearth		Oxygen		Electric	
	1974	*1980*	*1974*	*1980*	*1974*	*1980*
Bulgaria	23.4	19.0	57.8	55.5	18.8	25.5
East Germany	71.5	62.9	4.1	9.3	20.8	26.4
Hungary	90.5	90.7	—	—	9.5	9.3
Poland	67.7	46.6	22.8	39.0	9.5	14.4
Rumania	52.9	36.7	36.2	44.0	10.8	19.3

Source: International Iron and Steel Institute. *World Steel Figures 1981* (Brussels), p. 7.

and 8.1 percent of their respective outputs. By 1980, there was some change as the Soviets increased to 12.5 percent, Poland to 4.4 percent, Czechoslovakia to 0.7 percent, East Germany to 11 percent, and Hungary to 35 percent.[3]

The Soviet steel industry has made notable progress in terms of coke-oven and blast-furnace operations. It was the pioneer in the dry quenching of coke, the process the Japanese are now adopting. Further, its blast furnaces in some locations are capable of producing 10,000 tons a day, which is the equivalent of the Japanese operations as well as the two new furnaces in the United States. In the 1960s, the Russians ran their blast furnaces on a 100-percent sinter charge and in this they led the field. In terms of rolling mills, the Soviets leave much to be desired. They have quality problems with items like cold-rolled sheets, which they are forced to import for the manufacture of automobiles. They also import considerable quantities of large-diameter pipe, as well as plates.

Raw Materials

There is an abundance of iron ore in the USSR, so much so that the USSR leads the world in iron-ore production by a considerable margin. In 1980, it reached a high point of 244.8 million tons, which was more than twice as much as Australia, the second largest producer in 1980. The ore, however, is relatively low in iron content compared with that of Australia and Brazil. In 1950, the average iron content was approximately 50 percent; by 1960, it had fallen to 44.5 percent; in 1970, it was 37.3 percent; and in 1980, approximately 35 percent. As a consequence, the amount of ore that had to be beneficiated increased from 37 percent of the total in 1950 to about 87 percent in 1980.

The Eastern bloc Communist countries are deficient in iron ore and, therefore, receive most of their requirements from the USSR. For example, Poland imported 17.2 million tons of iron ore and concentrates in 1978; of this amount, 11.6 million tons came from the USSR, 3.2 million from Sweden, and 1.9 million from Brazil. Czechoslovakia imported 15.6 million tons of iron ore and concentrates in 1978, of which 13.1 million came from the USSR, 1 million from Brazil, and 0.5 million from Sweden. Rumania imported 13.8 million tons of ore and concentrates in 1978, of which the USSR provided 4.4 million, Brazil 0.9 million, and Spain 0.7 million tons. Total iron-ore exports from the USSR to all countries in 1978 was 40.6 million tons, placing it third on the list of exporting countries, behind Australia and Brazil. In 1979 and 1980, however, it was surpassed by Canada.

In terms of coal, the USSR has large supplies, but it is not the highest grade for metallurgical use and, therefore, a significant amount of Polish low-volatile coal must be imported to improve the blend of coals for the production of coke.

Poland has an extensive supply of good coking coal and it exports a very large tonnage annually. In 1978, the exports amounted to 40.1 million tons, of which 9.9 million went to the USSR, 4.8 million to France, 4.0 million to Finland, and 3.4 million to Italy.

Czechoslovakia has substantial coal reserves and much of its bituminous coal production is of metallurgical quality. It is stated "deep coal mining of bituminous coal in Czechoslovakia meets demand for coking coal for the domestic industry as well as for export."[4] It is interesting to note, however, that in spite of these claims, Czechoslovakia imported 5.6 million tons of coal in 1978, 3.2 million from the USSR and 2.4 million from Poland.

In addition to the iron-ore imports, Rumania must look to the outside for a certain amount of its coal. In 1978, it imported 4.7 million tons. Currently, it is contracting for large imports of coke from the United Kingdom in order to provide sufficient blast-furnace fuel.

Trade

The USSR conducts a large trade in steel, having exported an average of 7.0 million tons annually for the past decade. Imports have grown substantially during the same period, rising from 2.9 million tons in 1970 to 8.5 million tons in 1979, with an estimated 8.0 million tons in 1980.

The USSR imports those products that it cannot satisfactorily produce at home. These include cold-rolled sheets from Western Europe to supply the needs of its automotive industry, as well as plates and large-diameter pipe from Japan and Western Europe.

In terms of exports, the USSR supplies steel to the Eastern bloc countries and some to the Third World. One of the largest recipients of Soviet steel is East Germany, which has consistently imported a total of more than 4.0 million tons of foreign steel for the last six years.

The remainder of the Communist bloc conducts a significant steel trade in terms of imports and exports.

Czechoslovakia is the leading exporter, with an average of 3.5 million tons during the past five years. Its imports are substantially lower, averaging approximately 600 thousand tons in the last three years.

Poland has maintained a steady export figure of about 1.5 million tons for most of the past decade. This rose to almost 2.0 million tons in 1978 and 1979. Poland's imports reached a high of 3.8 million tons in 1975; however, in the past few years imports have averaged about 2.0 million tons.

Rumania is an active exporter of steel, having exported 2.0 million tons in the last few years, a significant portion of which went to the Western world.

Future Prospects

The USSR, as well as the Eastern bloc, have centrally planned and controlled economies with five-year plans. The USSR had set objectives in its tenth five-year plan, covering 1976 through 1980, for steel production to rise by 1980 to 160–170 million tons. This objective was not realized and consequently revised downward to 157 million tons.[5] Production in 1980 of 148 million tons was significantly below the objective. A number of reasons are given for this failure to fulfill targets for steel production. It is said that "The poor performance of the ferrous metallurgy industries resulted mainly from delays in plant modernization, labor shortages, and transportation problems . . . . The balance in the development of ferrous metallurgy works is not being maintained; for instance, at Cherepovets a powerful rolling mill was built before the blast furnace and oxygen-converter shop, and thus metal had to be brought from Zhadanov and other distant places."[6] It is further stated, based on Soviet documentation, that "more than 50 percent of the heavy and labor-intensive operations at ferrous metallurgy enterprises have to be done manually. Over 67 percent of the workers in blast furnace production perform manual operations. The situation is somewhat better in steel smelting [sic] production, where 55.8 percent of the workers perform manual labor."[7]

As of mid 1982, the USSR is still very much concerned with improving its steel industry. Raw-steel production is to be increased significantly by

1985. However, special attention will be devoted to improving the quality of steel and expanding the range of products. The USSR plans to increase the percentage of steel produced in the oxygen converter as well as the electric furnace. In a number of locations, oxygen converters are under construction and it appears that by 1985 some 35 percent of the output will come from the oxygen process and possibly 17 or 18 percent from the electric furnace, thus reducing the open-hearth contribution to less than 50 percent. A recent announcement indicates that two of the USSR's major steel works will be rebuilt by 1990. These are the country's largest mills at Magnitogorsk and Kuznetsk. Both plants have substantial open-hearth capacity, which will be replaced by oxygen converters and electric-arc furnaces.[8] The plans for the next five years also include the construction of mimimills (which are called Kompact mills). The first of these will have two electric furnaces of 100-ton capacity each, capable of producing a total of 550,000 tons of raw steel. There will also be two billet casters and a bar mill with a 500,000 ton capacity. In addition to this first mill, depending on its success, others will be built, perhaps as many as four or five in the next five to seven years.

The Soviets have contracted with Korf Industries to construct facilities for 2 million tons of directly reduced iron. The units are under construction at Kursk. Some of the output will be consumed there and the remainder will be shipped to other plants in the USSR. Production of DRI is scheduled to start in 1983.

Of the Eastern bloc countries, Poland is by far the largest steel producer, with an output of 19.5 million tons in 1980 as compared with 14.9 million tons for Czechoslovakia and 13.2 million for Rumania. Poland had elaborate plans to increase its raw-steel production to 25 million tons by 1985, which was to be achieved, in part, by the modernization of some units at the Lenin steel plant in Krakow (which produced 6.7 million tons of raw steel in 1979) and the expansion of the new plant at Katowice. The Katowice complex produced 4.5 million tons of steel in 1979 and was scheduled to reach 9 million tons. Recent political disturbances in Poland have made it very difficult to achieve this goal. In fact, steel production fell to 15.6 million tons in 1981.

Czechoslovakian expansion plans are relatively modest, calling for 16.3 million tons of steel by 1985 and 20.4 million tons by 1990. This will take place through the renovation of a number of existing steel works, which will result in the replacement of open-hearth capacity.

Rumania has the most ambitious plans for expansion, in spite of the fact that it must import very large amounts of ore and considerable amounts of coal and coke. It was hoped that some 15.6 million tons of crude steel could be produced in 1980. The output fell about 2.5 million tons short of this goal. By 1990, the industry in Rumania expects to be producing 25 million tons, with a significant amount of this designated for export. The past performance of Rumanian steel expansion resulted in a doubling of production between 1971 and 1980. The future goal which calls for doubling production in the 1980s, will be more difficult to achieve.

East Germany has witnessed a relatively small growth during the past decade. Steel production was 5.7 million tons in 1971 and 7.3 million in 1980. Currently, it has plans to install approximately 2 million tons of steel-making capacity at Eisenhuttenkombinat Ost Works. As of mid 1982, this plant has an annual iron-production capability of approximately 2 million tons. The iron is shipped to the USSR and processed into steel and hot-rolled coils, which are shipped back to the East German plant to be rolled into cold-rolled sheets on a cold-reduction mill that has a capacity of more than 1 million tons. The installation of the basic-oxygen converters and continous casting will allow the plant to ship slabs to the USSR to be rolled into hot-rolled bands, returned to the mill, and converted into cold-rolled sheets. The hot-rolled bands, which currently come back to Germany, account, in part, for the large tonnage of imports into East Germany. In 1979, they amounted to 4.2 million tons, dropping slightly to 4 million tons in 1980. Among the Eastern bloc countries, exclusive of the USSR, East Germany is by far the largest importer of steel. Hungary and Bulgaria, in 1981, produced 3.6 and 2.6 million tons.

The USSR has been active in selling its technology abroad. It has agreed to assist in the construction of a 3-million-ton steel plant in India at Vishakapatnam. It is also involved in the construction of a 1.5-million-ton steel works in Nigeria, as well as in rendering assistance in the construction of a steel works in Algeria.

Notes

1. International Iron and Steel Institute. *Steel Statistical Yearbook 1981* (Brussels), p. 2.

2. Ibid.

3. Ibid., p. 7.

4. *Minerals Yearbook, Vol. III* (Washington, D.C.: U.S. Department of the Interior, Bureau of Mines, 1978–1979), p. 303.

5. Ibid., p. 988.

6. Ibid., p. 957.

7. *Socialist Industry* (Moscow: June 5, 1979), p. 1. Quoted in *Minerals Yearbook, Vol. III* (Washington, D.C.: Department of the Interior, Bureau of Mines).

8. *Metal Bulletin* (London: Metal Bulletin P.L.C., April 2, 1982), p. 5.

9 International Trade

International trade in the steel industry has changed radically since World War II due to a number of developments, which include:

1. An unprecedented fivefold growth in global steel production over a twenty-six-year period, which provided excess steel for export over domestic needs.
2. A shift from a predominantly privately owned industry to one where a majority of the industry is either government owned or heavily subsidized, resulting in the greater use of the steel industry as a tool for government policy.
3. The heavy emphasis that government policies place on employment as a prime consideration. Consequently, to maintain employment in government-owned or subsidized mills, steel has often been overproduced and underpriced in the international market.
4. The dominance of social, political, and production considerations rather than economics in determining many decisions on capacity.
5. The development of an extensive trade between steel-producing countries, resulting at times in severe competition for each other's markets.
6. The rise of Japan from a position of insignificance in world trade to that of dominant leader.
7. The change in position of the United States from an exporting nation in the prewar and immediate postwar periods to the largest net importer of steel. As a consequence, its market has at times become a battleground for world exporters.
8. The development of various protectionist devices established by many countries throughout the world, relating specifically to steel.
9. The growth of a very large trade in iron ore, much of it between the Third World and the industrialized countries.
10. The installation of steelmaking capacity in the Third World and the subsequent reduction of Third World steel markets to industrialized-world exporters.

In the immediate postwar years up to 1950, steel output was limited since most of the industrial countries involved in World War II suffered severe damage to their steel-producing facilities. The relatively small production of most countries was sorely needed at home for a massive rebuilding effort, so very little was available for export.

In 1947, the first full year of production after the war, worldwide output of raw steel was 136 million tons, of which 77 million, or 57 percent, was produced in the United States. In that year, the United States exported 4.2 million tons of finished steel products, which was by far the largest amount moved into the international market by any producer. Export tonnage was limited by the low levels of production since only two other countries produced more than 10 million tons of raw steel, the United Kingdom with 12.9 million tons, and the USSR with an estimated 14.7 million tons. Production from the rest of the steel-producing nations was minimal. In Japan it was 946 thousand tons, in France 5.8 million, in Belgium 2.9 million, and in West Germany 3.7 million tons. In these countries, finished steel shipments to domestic consumers represented approximately 70 percent of their raw-steel output, leaving very little for export.

World production increased significantly between 1947 and 1950, moving up to 192 million tons of raw steel, an increase of almost 40 percent in the three-year period. At this point, a number of countries began to export some steel, so that 20.5 million tons of raw-steel equivalent, or slightly more than 10 percent of total world production, was shipped into the export market. (Raw-steel equivalent is the amount of raw steel necessary to produce the finished products that are ultimately shipped to the market.)

By 1960, 15.3 percent of the world's steel production, or some 57.2 million tons of raw-steel equivalent, was moved in international trade. During the next ten years, this figure more than doubled as it reached 117.5 million tons in 1970, or 19.7 percent of world production. By this time, Japan and Western Europe had become significant exporters of steel, while the U.S. participation in world trade had diminished sharply with the exception of one or two years.

During the 1960s and 1970s, the Western European and Japanese industries were very much export oriented, for in both instances the expansion of capacity provided much more steel-producing potential than was needed to satisfy domestic demand. Up to 1974, with a continual growth in world consumption, the Western European countries and Japan showed consistent increases in exports.

Table 9-1 gives exports for the EEC countries, Japan, and the United States from 1967 to 1974, and for most countries indicates an almost-constant increase. The increases in export tonnage during those seven years range from 60 percent to well over 100 percent, depending on the country. In the EEC, West Germany showed the largest tonnage increase, while in percentage terms, exports from both Italy and Holland more than doubled. The United Kingdom was an exception, as exports remained relatively stable.

During the same period, Japanese exports increased from 8.8 million to 33.1 million tons, while U.S. participation in the world market presented an erratic pattern as the table indicates. The U.S. figures showed substantial

Table 9–1
Exports for the European Common Market Countries, Japan, and the United States: 1967–1974
(millions of metric tons)

Year	Belgium-Luxembourg	West Germany	France	Italy	Netherlands	United Kingdom	Denmark	Japan	United States
1967	9.6	12.0	6.4	2.1	2.4	3.9	0.1	9.9	1.6
1968	11.2	12.8	6.9	2.5	2.6	4.4	0.3	13.2	2.0
1969	12.8	12.6	6.7	1.9	3.2	3.9	0.2	16.6	4.8
1970	12.8	11.9	7.5	1.8	3.1	4.0	0.3	18.0	6.5
1971	12.4	13.1	7.9	3.1	3.8	4.8	0.3	24.7	2.6
1972	14.4	13.8	8.4	3.8	4.4	4.6	0.3	22.0	2.7
1973	16.0	17.1	8.6	3.5	4.8	4.2	0.3	25.6	3.7
1974	16.6	22.2	10.3	4.8	5.2	3.3	0.3	33.1	5.3

Source: *Steel Statistical Yearbook 1980* (Brussels, International Iron and Steel Institute), p. 24.
Note: Exports of the EEC countries include those exchanged among its members.

fluctuations with the 6.5 million tons in 1970 registering an all-time high, which has not since been equaled.

Third World countries increased their imports of steel significantly from 1967 to 1974. This was particularly true of those in Latin America, such as Argentina, Brazil, and Venezuela, as indicated in table 9-2. Other Third World countries, such as Indonesia, India, and South Korea, also recorded sharp increases in imports, while the Middle East as a whole almost quadrupled its imports. Most of these countries needed steel, so there was no question of resistance to imports.

The main exception to the need for imports was the United States, which nevertheless watched its imports grow dramatically from 3.0 million tons of products in 1960 to 16.3 million in 1968. The growth in the European and Japanese industries and the increase in their exports to the United States constrained the growth of the U.S. steel industry as table 9-3 indicates. It presents indexes for steel production from 1964 to 1974 for the United States, Japan, and the EEC countries. The U.S. industry showed very little growth during the period, with a major setback from 1969 to 1971, following a lesser but significant drop in 1967. The Japanese and Europeans, however, recorded almost constant growth, with only some minor interruptions.

U.S. imports rose sharply during the 1960s for a number of reasons, including:

1. The growth in steel capacity in other parts of the world, providing excess tonnnage for export.
2. Low costs of production due to new mills and low wage rates.
3. Low prices offered by foreign producers to obtain a foothold in the U.S. market.
4. The triennial labor negotiations, which prompted steel consumers to hedge against a possible strike by increasing inventories with large orders not only from the domestic suppliers in the months before the strike deadline, but also from foreign suppliers who were committed to deliver both before and after the strike deadline.

In 1959, a four-month steel strike left steel consumers short of material, and although they struggled to obtain it wherever possible, they could not get enough. Mindful of this, steel consumers watched the labor negotiations in 1965 with an anxious eye. Determined not to be caught short, they bought steel from a number of foreign sources and imports increased that year by approximately 3.6 million tons, from 5.8 million in 1964 to 9.4 million in 1965. In the two subsequent nonnegotiating years, imports did not drop back to the 1964 level, but were sustained at a level slightly above the 1965 record. In 1968, another negotiating year, in order to protect themselves, consumers accumulated inventory by buying from foreign sources to

Table 9-2
Imports of Steel for Selected Third World Countries: 1967–1974
(millions of metric tons)

Year	Argentina	Brazil	Mexico	Venezuela	Iran	Saudi Arabia	China	India	Indonesia	Republic of South Korea
1967	0.5	0.3	0.2	0.4	1.0	0.1	1.6	0.2[a]	0.1	0.3
1968	0.4	0.3	0.2	0.4	1.3	0.2	1.7	0.5	0.2	0.4
1969	0.7	0.5	0.2	0.5	1.2	0.2	1.8	0.4	0.3	0.6
1970	0.8	0.6	0.2	0.5	0.9	0.2	2.3	0.6	0.4	0.6
1971	1.6	1.2	0.2	0.6	1.6	0.3	3.0	1.2	0.5	0.8
1972	1.6	1.1	0.2	0.7	1.3	0.4	2.3	1.3	0.8	1.1
1973	1.9	1.8	0.3	0.8	1.9	0.6	3.7	1.0	1.1	2.1
1974	1.5	4.2	0.5	1.1	2.3	0.7	3.6	1.2	1.1	2.2

Source: *Statistical Reports* (Brussels: International Iron and Steel Institute, various years.)
[a]Represents six months of imports.

Table 9-3
Indexes for Steel Production for the United States, Japan, and the EEC Countries: 1964–1974
(1974 = 100)

Country	1964	1965	1966	1967	1968	1969	1970	1971	1972	1973	1974
Belgium	53.8	56.5	54.9	59.9	71.3	79.1	77.7	76.9	89.6	95.7	100.0
France	73.2	72.6	72.5	72.7	75.5	83.3	88.0	84.6	89.2	93.5	100.0
West Germany	70.1	69.2	66.3	69.0	77.3	85.1	84.6	75.7	82.1	93.0	100.0
Italy	41.1	53.3	57.3	66.7	71.3	69.0	72.6	73.3	83.2	88.2	100.0
Luxembourg	70.7	71.1	68.1	69.5	75.0	85.6	84.7	81.3	84.7	91.9	100.0
Netherlands	45.6	53.9	55.8	58.3	63.5	80.9	86.2	87.1	95.4	96.3	100.0
Japan	34.0	35.1	40.8	53.1	57.2	70.1	79.7	75.6	82.7	101.9	100.0
United States	87.2	90.2	92.0	87.3	90.2	96.9	90.3	82.7	91.3	103.2	100.0

Source: Indexes are based on production statistics for the years involved, taken from Japan Iron and Steel Federation Monthly Bulletin, p. 1151. Statistical reports and annual statistical reports of American Iron and Steel Institute.

such an extent the imports reached 16.3 million tons. This represented 16.7 percent of apparent consumption and brought an urgent call for immediate action to restrain imports. The rapid growth of imports and the relatively stable exports are indicated in the following list, which provides figures (in millions of tons) for both from 1960 to 1969.[1]

	Imports	Exports
1960	3.0	2.7
1961	2.9	1.8
1962	3.7	1.8
1963	4.9	2.0
1964	5.8	3.1
1965	9.4	2.3
1966	9.8	1.6
1967	10.4	1.5
1968	16.3	2.0
1969	12.7	4.7

In the two years prior to 1968, as imports were sustained at a high level, a movement for some type of protection against foreign steel producers was developed. At first the suggestion was made that a temporary tariff be put into effect to halt what were termed unfair and unrestricted imports. This was intended to give the U.S. steel industry time to modernize its plants and equipment and become competitive with the rest of the world. Subsequently, a bill was introduced in the U.S. Senate to put a quota on imports, limiting them to 9.6 percent of the U.S. market. Hearings on the subject witnessed a joint appearance of the chairman of the American Iron and Steel Institute, as well as the president of the United Steelworkers of America, in support of restricted imports. There was opposition to the quota from steel importers, who would have been adversely affected, and from a number of industrialists, as well as bankers whose companies carried on a large volume of international business.

Rather than await the outcome of the hearings and the possible establishment of quotas by law, the Germans and the Japanese made a sudden, dramatic move in the summer of 1968 by proposing to congressional leaders the possibility of placing voluntary restrictions on their steel exports to the United States. No total tonnage was set forth but a growth rate of 7 percent a year was suggested by the Japanese. This caused much concern among the U.S. steelmen, who could envision the growth of the U.S. steel market being absorbed year after year by foreign producers.

After prolonged negotiations between the Western Europeans and Japanese steelmakers, as well as some U.S. government officials and private parties, a total figure of 12.7 million tons for 1969 U.S. imports was established. This was a compromise between the level put forth by the Western Europeans who had suggested 14.5 million tons and the Japanese who had proposed 11.0 million tons.

The Japanese and the EEC were each allotted 41 percent of the total, leaving 18 percent for the rest of the world. The arrangement, which was to last for three years, stipulated that the growth in steel exports to the United States was to be limited to 5 percent in each of the two following years.

Two important exporters of steel to the United States, Canada and the United Kingdom, did not participate in the agreement and, therefore, were not obliged to follow its conditions. Likewise, all other steel-producing countries in the world were not included and thus technically were not under any restraints.

In 1969, total U.S. imports fell to 12.7 million tons due partly to the voluntary restraint agreements, as well as to an improvement in the world steel market. The year 1970 witnessed a further decline in imports, as they dropped to 12.1 million tons, which was actually 8 percent less than permitted by the voluntary restraint agreements. This cut in imports was the result of increased steel consumption elsewhere in the world.

It is interesting to note that the revenue paid out for the 12.1 million tons of imports in 1970 was almost identical with that paid for 16.3 million tons of imports in 1968. In the earlier year, the imports were valued at $1.976 billion and, in 1970, with a reduction of over 4.2 million tons, at $1.967 billion. Some of this was due to a shift to the more expensive steel products. However, much of it was due to the decreased tonnage, which allowed the importers to charge higher prices.

In 1971, the third and last year of the voluntary restraint agreements, negotiations were reopened for a three-year extension. In the midst of these negotiations, the Nixon administration imposed a 10 percent surcharge on all imports into the United States as of August 15, 1971. The Europeans argued that this violated the quota agreement and proceeded to increase their exports, as did the Japanese, with the result that 1971 witnessed a record 16.6 million tons of imports, 2.5 million tons above that allowed by the voluntary restraint agreements. Negotiations to extend the agreements dragged on until May 1972, principally because the Europeans could not come to terms. The Japanese reached an accord considerably before the Europeans, which was not surprising since the Europeans represented seven nations with somewhat differing points of view. The agreement was short-lived.

In 1972, a suit by the Consumers Union of the United States was filed against the U.S. government, as well as some European and Japanese steel companies, claiming that the voluntary restraint agreements were in violation of the Sherman Anti-Trust Act. This suit, with possible triple damages if it were decided in favor of the Consumers Union, and the increase in business in 1973 when steel was in great demand all over the world, brought an end to the voluntary restraint agreements.

In the euphoria of the worldwide steel boom during 1973 and 1974, all thoughts of restraining imports were forgotten, not only in the United States but virtually everywhere. Imports no longer came into a country below domestic prices but at figures far above domestic prices, since the intense demand in relation to supply gave rise to a much higher price level. Domestic prices were controlled in many countries throughout the world; however, export prices were not and a number of companies took advantage of the export trade to increase their revenues. A major Western European steel producer sold plates to the United States in January 1972 at $119 per metric ton FOB the Western European mill. By July 1974, the price had risen to $440. During the two boom years, there were practically no problems in relation to trade; the only problem was in obtaining enough steel.

In 1975, the boom came to an abrupt end in the industrialized world as demand and, consequently, production suffered severe cutbacks. This was particularly so in the export trade. Total exports of the EEC (a significant portion of which were exchanged between members) fell substantially from an all-time high of 62.8 million tons in 1974 to 51.2 million tons in 1975. Japan's exports dropped from 33.1 million in 1974 to 30.0 million tons in 1975. The United States, which exported a substantial tonnage in relation to its previous practice, watched this fall from 5.3 million tons in 1974 to 2.7 million in 1975.

The decline in trade was a blow to the Western Europeans and the Japanese since they had come to depend heavily on the export markets to sustain their operations. In 1974, EEC members shipped approximately 29 million tons outside of the EEC. The principal recipients were the United States, Latin America, Eastern Europe, and the Middle East. This 1974 all-time high of total EEC exports both internal and external, represented an abrupt increase of almost 13 million tons over 1972, indicating the extent of the world boom. In the same years, Japanese exports advanced by over 11 million tons.

In contrast, 1974 exports from the United States rose only 2.6 million tons over 1972, indicating the limited participation in the export market by U.S. producers. This was not due to a lack of interest but rather a desire to serve the U.S. market, which was clamoring for steel. The U.S. producers had every opportunity to sell steel in the export market at much higher prices than were available domestically, since price controls were in force up until April 30, 1974. For the most part, however, they did not desert their customers who had remained with them during the leaner days of the late

1960s, principally because they did not wish the customers to desert them at some future time when imports might again be available at lower prices.

Japanese Export Activity

During the 1970s, the Japanese, in addition to their exports to the United States, became more active in other market areas, some of which were traditionally served by the EEC producers. For example, Japanese exports to the Middle East increased from 576,000 tons in 1970 to 3.2 million tons in 1974. In Latin America, Japanese exports rose from 1.6 million tons in 1970 to 4.4 million in 1974; in Africa, the increase was from 830,000 tons in 1970 to 1.4 million in 1974.

In several of the Western European countries outside of the EEC, there were also significant Japanese increases; in Greece, exports doubled from 250,000 tons in 1970 to 506,000 tons in 1974. A significant increase was evident in Japanese sales to the Soviet Union, which rose from 228,000 tons to 1.3 million over the same period. Spain showed erratic behavior as an importer from Japan, accounting for 158,000 tons in 1970, dropping to well below 100,000 in the next four years, and then suddenly increasing to 831,000 tons in 1975 and 887,000 tons in 1976.

The Japanese had also penetrated the EEC in the early 1970s with almost 2 million tons in 1971. This was double the previous year's tonnage and brought immediate action from the EEC. An agreement between the Ministry of International Trade (MITI in Japan) and the EEC resulted in the establishment of a steel-export cartel, whose purpose it was to limit and monitor shipments of the six major Japanese steel producers to the EEC. Quotas was set up permitting the total shipment of 1.15 million metric tons of products. This was limited to one year (1972) and had a salutory effect for the Europeans, as Japanese exports to the EEC in that year dropped to 1.5 million tons. The agreement was extended for two additional years with an increase in the quota to 1.22 million metric tons. In 1973, imports from Japan fell to 1.2 million tons and, in 1974, dropped further to 1.0 million tons. With the steel prosperity of 1974 and the relatively low tonnage of Japanese imports, the Europeans agreed to discontinue the quota and the Japanese cartel was disbanded.

The Western Europeans came into conflict with the Japanese in various market areas in South America, Africa, the Middle East, and in non-EEC countries in Europe. This was expected competition, and during 1973 and 1974 there was plenty of business for all. As the boom came to an end in 1975, conditions changed significantly. Of prime importance to the EEC producers was the fact that the Japanese took advantage of the cancellations of the quotas and greatly increased their imports to the EEC in early 1975. During the first six months, tonnage amounted to more than was shipped in during the entire year of 1974; the total for 1975 was 1.6 million tons. This brought swift action again and the 1.22 million-ton quota was reinstituted in 1976, although it did not have its full effect until 1977.

The figures for Japanese imports into the EEC for 1976 and 1977 are significantly higher than the quota previously stated; however, it should be remembered that the quota applied only to the six integrated producers. A very substantial amount of tonnage is produced by Japanese electric-furnace operators, and a significant amount of this found its way into the EEC.

In 1976, the Western Europeans withdrew to some extent from the export market, since their domestic prices were more attractive than those that could be obtained abroad. Shipments to the United States from the EEC fell to 2.9 million tons, which was half of the 1974 figure. This left a void that was promptly filled by the Japanese, who increased total exports to record heights in that year, as they topped 37 million tons. Shipments to the United States were 7.2 million tons (almost 2 million tons above 1974). By 1977, with low domestic prices, the Europeans came back into their traditional export markets and put on an aggressive campaign to regain what they had lost. As a result, in 1977, with export tonnage throughout the world at virtually the same level as in the previous year, the EEC countries recovered 2.7 million tons, while the Japanese lost approximately 2.0 million tons. The U.S. market, which was at that time and continues to be the largest and most lucrative in the world, became a battleground.

The 1977 Crisis

In 1977, imports to the United States rose to 17.5 million tons from 13.0 million in the previous year, an increase of 4.5 million tons. In the struggle for shares of the U.S. market, the Japanese shipped 7.1 million tons as compared with 6.2 million from the EEC countries. Canada participated in the U.S. market with 1.7 million tons.

The price structure was reduced to a shambles as imports came in at reductions of up to $100 below the U.S. list price. As a consequence, the U.S. steel industry's profits on domestic shipments of 82.7 million tons fell sharply. On sales of almost $40 billion, the American Iron and Steel Institute reported profits for the industry of $22.3 million, or about one-twentieth of one percent of sales. This figure was also affected by losses due to plant closures. However, it did not include the losses of the Alan Wood Steel Company, which closed its doors in August 1977.

The year 1977, which can only be described as disastrous for the United States in terms of profits, witnessed the permanent closing of 5.2 million tons of U.S. steel capacity, including:

1. Alan Wood Steel Company, an operation with a 150-year history.
2. Parts of Bethlehem Steel Corporation's Lackawanna and Johnstown plants.
3. The Youngstown Sheet and Tube Company operations in Youngstown, Ohio.

One does not have to look far for an explanation to determine why 1977, which was the eighth best year in terms of U.S. steel shipments, was the worst postwar year in terms of profits. Obviously, a lack of volume was not the answer. The explanation was to be found in increasing costs and the deterioration of the price structure.

These financial problems, which had been developing for a number of years, culminated not only in the permanent closure of the aforementioned plants but also in the permanent loss of 20,000 jobs. Nevertheless, the layoffs at Alan Wood and Bethlehem Steel did little to produce a reaction on the part of the administration in Washington. However, when Youngstown Sheet and Tube announced the closing of its facilities in the Youngstown area with the accompanying loss of 5,000 jobs, the formation of a steel caucus in the U.S. Congress, as well as a coalition of mayors representing the steel cities involved, produced action. It should be pointed out that in addition to the loss of the raw-steel capacity, a number of steel companies closed various types of facilities and raw-material operations, adding to the permanent loss of jobs.

Prior to the September 1977 announcement of the Youngstown Sheet and Tube closing, the Gilmore Steel Company in Oregon instituted an anti-dumping suit against the Japanese who were charged with selling plates on the West Coast, in violation of the trade laws. The case was decided in the first instance in favor of Gilmore. The finding indicated that Japanese plates were being sold at 32 percent below the required price and this put a temporary halt to the importation of this product from Japan to the West Coast of the United States. In a subsequent review, the percentage was decreased from 32 percent but varied for the individual companies involved.

As a result of the first decision, the Europeans and Japanese steelmakers held a meeting in Rome on October 10, 1977, at which they decided to restrict their exports to the United States based on a voluntary quota agreement, a device that had been used from 1969 through 1973. The quota plan was termed unacceptable by the administration in Washington, as well as by some U.S. steel-industry spokesmen.

To meet the crisis, a White House meeting was held on October 13, 1977, to determine a course of action. A steel task force was appointed to quickly study the problem and suggest remedies. In early December, a document titled "A Comprehensive Program for the Steel Industry" was sent to President Carter. The program made several recommendations, the principal one being the institution of a trigger-price mechanism, in which it was prescribed "that the Department of the Treasury, in administering the Antidumping Act, set up a system of trigger prices, based on the full cost of production, including appropriate capital charges, of steel-mill products by the most efficient foreign steel producers (then considered to be the Japanese), which would be used as a basis for monitoring imports of steel into the United States and for initiating accelerated antidumping investigations with respect to imports priced below the trigger prices." This was designed to provide "a basis for initiating antidumping investigations without any prior industry complaint."[2] The trigger prices were to be based on cost data provided by Japanese producers plus freight to the United States plus a minimal profit, and were to be adjusted quarterly. It was conceded that "the implementation of the trigger-price mechanism may not prevent less efficient producers from selling steel products at less than 'fair values' within the meaning of the Antidumping Act."[3]

During the preparation of the comprehensive program for the steel industry, pressure to head off the pending antidumping suits were exerted at the highest levels of the U.S. government by the Europeans and the Japanese. Both were very much in favor of any accommodation to avoid enforcement of the Antidumping Act. Thus, once the trigger-price mechanism was proposed, they accepted it readily.

Other aspects of the program included suggestions to adjust depreciation write-offs for steel equipment, improve the relationship between the steel industry and the Environmental Protection Agency, assist communities where steelworkers lost jobs, and provide financial assistance in the form of loan guarantees wherever the application was judged worthy. Unfortunately, all of these provisions were never fully carried out.

It was privately expressed that the trigger price should reduce annual imports to less than 11 million metric tons. A delay in putting the program in force for some four months, up until May 1978, was announced in order to give America's trading partners an opportunity to process and ship steel that had been ordered. This resulted in an unprecedented influx of 7.2 million tons of imports in the first four months of 1978, which helped bring the total for that year to a record 19.2 million metric tons.

The trigger-price mechanism had the effect of raising prices on imports, which not only benefited the U.S. steel producers, but was of significant help to those exporting to the United States. In 1977, the U.S. market was a jungle, with deep price cutting prevalent. The low prices resulted in losses for everyone concerned, and many Europeans could sustain such losses only because they were either government owned or government subsidized. The chairman of one of the large European companies, when asked why he was so adamant about shipping steel to the United States at a loss of $50 to $70 per ton, said his prime worry was unemployment at home, which in effect was being exported to the United States. The advent of the trigger price restored some semblance of order to the U.S. market and became an integral part of the sales policy of importers, who very often hoped for quarterly increases so that they could benefit from them.

In 1978 and 1979, the steel business in the United States improved considerably and, although there was a slight improvement in Europe, most companies there continued to sustain losses. In 1980, conditions on a worldwide basis deteriorated, and the United States Steel Corporation, unsatisfied with the trigger price as an operative device, filed antidumping suits against the Europeans, which prompted the U.S. government to suspend the trigger-price mechanism.

Conditions in the European market deteriorated rapidly in mid 1980 and losses reached staggering proportions, which made the antidumping suits a matter of international diplomacy. At a summit conference in Italy, President Carter was asked to have them withdrawn and pressure was put on the United States Steel Corporation to withdraw its suits in favor of reinstating an improved trigger-price mechanism. This was finally agreed to in October when the suits were dropped and the trigger-price system restored.

The worsening conditions during 1980 and 1981 in Western Europe, and the plans to remedy them, have been described earlier. How successfully these can be carried out remains to be seen. One thing is certain, the results will have a definite influence on worldwide steel trade.

In 1981, the Europeans appealed to the U.S. government to adjust the trigger-price mechanism to allow them greater access to the market. They maintained that the trigger price was above the actual domestic-price level in the United States, making it extremely difficult for importers to sell steel. They also claimed that their devalued currencies reduced their costs in terms of U.S. dollars, allowing them to sell below trigger prices. Many companies openly challenged the trigger price by shipping steel at a figure well below the trigger price, while others asked for preclearance, which is, in effect, permission to undersell the trigger price once it can be proven that the seller's costs are low enough to allow this.

The Commerce Department, whose function it is to administer the trigger price, evaluated the requests for preclearance that were submitted by several Western European steel producers and decided to deny them. In November 1981, the department brought action against five countries—Brazil, Rumania, France, Belgium, and South Africa—for trade violations with respect to hot-rolled sheets and plates. These cases were approved by the International Trade Commission.

In spite of the actions taken by the Commerce Department, a number of Western European steel producers continued to ship steel into the United

States at prices considerably below the trigger price. As a consequence, a number of U.S. steel producers prepared to file antidumping suits. They were asked to withhold these until conversations could be held between the EEC and U.S. government officials in Brussels in December 1981. Secretary of Commerce Baldridge and Trade Representative Brock met with the EEC representatives in a fruitless attempt to bring about a solution. The talks broke down on a number of points, which indicated a lack of agreement among the steelmen in the EEC, despite their desire for a solution.

The Western Europeans agreed to forecast tonnage to the United States for each quarter but would specify no sanctions if the forecast tonnage was exceeded. Neither were they willing to make a commitment concerning sales of steel to outside corporations, established offshore by the U.S. importers to buy steel below the trigger price and then resell it in the United States at the trigger price. The discussions were also vague on product coverage and the duration of an agreement that would be made.

After the Brussels meetings, the U.S. companies felt that they had no alternative but to file antidumping suits, which they did in mid January 1982. Some ninety-two suits were filed before the International Trade Commission; the commission agreed to process thirty-eight of these. When this decision was made, the Commerce Department withdrew the five cases that it had previously filed, stating that it did not have the manpower to process all the cases.

On June 11, 1982, the Commerce Department issued its preliminary findings on the extent to which a number of foreign steel companies were subsidized and would be subject to countervailing duties. The assessment ran from nothing or a fraction of 1 percent for some West German companies and Hoogovens of the Netherlands to 40.4 percent for British Steel.

For British Steel structurals, plates, hot-rolled sheets, cold-rolled sheets, cold-finished carbon bars, and hot-rolled carbon bars were involved. The assessments against the Belgian producer, Cockerill Sambre, were 20.6 percent on structurals and 21.8 percent of plate and hot-rolled sheets. In France, Sacilor was assessed 30 percent on structurals and hot- and cold-rolled sheets, while Usinor was assessed to the extent of 20 percent on the same product. The judgment on Italsider in Italy was 18.3 percent on hot- and cold-rolled sheets, while Brazil was assessed 8.6 percent on plates. The other country involved was South Africa where ISCOR was assessed from 12.5 percent to 16.3 percent on the products mentioned above, as well as galvanized sheets.

These preliminary findings required that the importers post a bond or cash equal to the amount of the subsidy. Being preliminary, the findings were subject to revision before the Commerce Department made a final determination. Subsequent to that, the International Trade Commission was to determine whether or not injury had been sustained by the American steel producers. If so, the final assesments would then be paid. The affected members of the European Community protested vigorously and attempted to have the judgment set aside by negotiating a trade agreement with the Commerce Department, which in its initial form was rejected by the U.S. complaintants. Without question, on the basis of the preliminary findings, imports would have been reduced significantly, since their prices would have been above the U.S. level in many cases.

Future of Steel Trade

An examination of the factors affecting the future of steel trade during the remainder of the 1980s indicates that it is reasonably certain that:

1. Specific protective devices regulating steel imports will remain in force in most countries throughout the world unless an actual worldwide shortage of steel develops.
2. Japan, the largest exporter, must continue to export or watch its industry decline significantly in size.
3. The Western European industry, although it will shrink during the first half of the decade, will continue to export more efficiently produced steel products.
4. The U.S. market will continue to be the target of exports from all countries, whether industrialized or developing.
5. The Third World countries, with more steelmaking capacity, will offer increasing competition in the world market and will reduce their imports considerably.
6. China, which has the largest prospect for steel-industry development in the decade ahead, will provide a significant market for steel products, as well as a market for steel-mill equipment.
7. Trade between the Eastern European countries, exclusive of the USSR and the West will increase. The USSR for the first half of the decade, will remain a net importer of steel from the Western world. In the second half of the decade, it could well enter into international trade outside of its own sphere.

Devices Restricting Trade

Despite many governments' desire to promote freer international trade, the need to maintain employment at home is a strong force in governmental decision making. The Western Europeans have established barriers against imports into the Common Market for Japanese steel, as well as that from other non-Common Market countries in Europe. As previously indicated, the first such protective action was taken in 1972. In 1974, with the world-wide boom in effect, restraints were dropped to be reinstated as soon as the boom ended. Without question, these restraints will remain in force as long as it is felt they are needed. Should another boom take place during the mid 1980s, restraints may be suspended but not removed as they were in 1974 because of the lesson gained from that experience.

The EEC restrictions placed on imports from countries other than Japan, including those in Western Europe, involve quotas controlling the amount of steel that each country can export to the EEC, as well as prices that must be charged. These are negotiated individually with each country.

The U.S. market has fewer restrictions on steel imports than almost any other country engaged in the manufacture of steel. It was protected in the past by (1) a voluntary restraint agreement that the Western Europeans and Japanese entered into in 1968 to restrict their exports to the United States, and (2) the trigger-price mechanism, which was installed in 1978 but also has been suspended. At present, the only protection that is available to steel producers in the United States is the procedure of filing antidumping and countervailing-duty suits.

The Japanese have no formal restrictions against imports. However, they have had insignificant tonnages of imports over the last decade in comparison with their export tonnage. In 1976, when they were exporting 37 million tons of finished steel product, imports amounted to a mere 176,000 tons; in fact, there was never a year between 1960 and 1978 when imports were as much as 500,000 tons. In 1979, they amounted to 1.6 million tons, which was a significant increase. The figures given by the Japanese for import tonnage are not in products but rather crude-steel equivalent, indicating that they have averaged less than 200,000 tons of products in most years. In fact, a statistical report issued by the Japanese Iron and Steel Federation does not even give a page or a table to steel imports. They are listed in a single column of one table on apparent steel consumption.

There was a substantial increase from 410,000 tons in 1978 to 1.6 million tons of raw-steel equivalent in 1979 and to 1.7 million tons in 1981.

In order to bring about a rational approach to international trade, the Japanese are particularly interested in developing the practice of orderly marketing. They have been long-time advocates of the procedure, which regulates the amount to steel traded among nations and also could have a definite stabilizing effect on price.

In the Third World, particularly where steelmaking facilities exist, imports are tightly controlled. In many instances, the government steel company sets the amounts, as well as the products that can be imported. There is no reason to think that this will change radically in the 1980s.

Steel protective devices are abhorred by most countries insofar as they limit their exports, although they employ these devices against imports.

They are regarded by most producing countries, at least in times of slack demand, as a necessary evil.

Japan Must Export

The Japanese steel industry, as previously indicated, is far too large to be sustained by its domestic market and, in fact, in the 1960s and early 1970s, its very large expansion was designed in great measure to serve the world market. Should its exports be cut back to any significant degree, that is, below 20 million tons, the industry would suffer a severe blow. Since 1974, the exports were constantly 30 million or more tons, reaching a high of 37 million tons in 1976 as a result of a vigorous campaign. Since then they have declined to an average of a little more than 30 million tons from 1978 to 1980. In 1981 exports dropped slightly to 29.1 million tons.

The Japanese have improved their efficiency so that they are profitable operating at approximately 70 percent of capacity. In 1980, production of 111.4 million tons on a basis of a 145-million ton capacity resulted in a 77 percent operating rate, which gave some of the largest integrated companies a significantly higher return on sales than had been previously achieved. However, should exports drop below the 20-million ton figure, this would reduce Japanese raw-steel production to a point well below 100 million tons, possibly 95 million tons. With the reduced tonnage, on the basis of a 145-million-ton capacity, this would amount to a 65 percent rate, at which level it would be extremely difficult for the companies to be profitable. Thus, the Japanese steel industry, in order to remain profitable will have to export more than 25 million tons. The question raised is whether or not this export rate can be sustained. In terms of shipments to consumer countries, in 1980, the major export markets were[4]

Country	Tonnage
United States	5,185,000
China	3,214,000
South Korea	1,829,000
Taiwan	1,778,000
Philippines	572,000
Middle East	4,036,000
EEC	621,000
Latin America	2,559,000
USSR	1,667,000
Indonesia	1,208,000
Other Far East	3,215,000

In examining these countries in terms of their possible future consumption of Japanese steel, the United States has been and will probably continue to be the largest consumer. Imports will continue at a significant rate of at least 4 million to 5 million tons a year in the foreseeable future. The markets are established and customer relations have been cemented over too many years to have them collapse. Japanese imports are in the United States to stay.

In terms of exports to China, the figures for 1980 and 1981 of 3.2 million tons and 2.1 million tons were substantially below the 1978 high of 5.6 million tons. The Chinese have retrenched to some extent, but this is not a permanent condition and it is expected that they will increase their imports from Japan in the years ahead. Tonnage could well reach 6 million and China might possibly surpass the United States as the number-one market. This is a bright spot in the Japanese export picture for the coming decade. As indicated, South Korea has recently expanded its steel-producing facilities and added a new wide, hot-strip mill. Since approximately 1 million tons of Japanese exports to Korea are in the form of strip-mill products, this market could be virtually closed to the Japanese. The new hot-strip mill has a capacity for 3 million tons and will provide South Korea with excess tonnage for exports, which will compete with the Japanese in the Far East. In fact, in 1981, about 1 million tons were sold in Japan.

Taiwan is installing new capacity that will bring its total production to more than 6 million tons, some of which will replace Japanese imports. The finishing facilities to process this steel consist of a hot-strip mill and a cold-strip mill with a million tons and 700,000 tons of capacity, respectively. Both are scheduled for completion in August 1982. Japan exports about 600,000 tons of hot-rolled sheets and coils to Taiwan; however, the completion of these two facilities will enable Taiwan not only to supply that which it now imports from Japan but also provide a surplus of more than 500,000 tons for export.

Indonesia, until it installs more steelmaking capacity and gears up to its direct-reduction units, will undoubtedly continue to import Japanese steel for some time to come. In 1974, import tonnage from Japan amounted to 728,000 tons, rising steadily to 1.2 million tons in 1980.

The Philippines imported a record 712,000 tons from Japan in 1978. However, the Filipinos have a plan to construct a 750,000 ton mill with Japanese assistance, which will certainly reduce Japanese imports. At present, the largest import item is hot-rolled coil, which is further processed at Illigan's cold-strip mill. Once basic-steel production is in operation in the Philippines, these imports will be substantially reduced, and the Philippines could well become a competitor of Japan in Southeast Asia.

The Middle East is a significant importer of Japanese steel, and although there are some small steel facilities recently erected and others under construction, it will continue to import large quantities of Japanese steel. The one possible exception could be Iran, if the program that was planned in the mid 1970s is put into operation. Political turmoil, which is reflected in the country's economy, must be settled before any projections can be made on future steel production in Iran.

Latin America has been a large market for the Japanese. However, since 1974, which was the peak year, this has dropped from 4.4 million tons to 2.6 million. New capacity in Brazil, Venezuela, Mexico, and Argentina will cut imports in the years ahead and the Japanese are bound to be hurt.

The USSR is a larger producer of steel than Japan. However, it imports considerable quantities of Japanese steel products. Recent negotiations have resulted in several hundred thousand tons of pipe and the possibility of 300,000 tons of plate. However, the Soviets are finding that the Western Europeans are willing to compete particularly on interest and credit terms and thus future Japanese participation in the Soviet market will depend on the willingness to respond to the Europeans.

There are indications that in a less-than-boom market, the Japanese will have to compete more vigorously for export business than they have in the past if they wish to maintain an export volume of 25 million or more tons. In addition to price, credit terms will figure prominently in this competition. Faced with these prospects, the Japanese industry, during the next five years, will concentrate, as it has in the past three years, on modernization and efficiency rather than increased capacity. The Japanese emphasis will be on producing lower cost and higher quality steel rather than more steel. This was emphasized in a recent statement by Yukata Takeda, president of Nippon Steel, who maintained that the modernization program "must be accomplished as soon as possible." This program, begun in 1977, was designed to strengthen Nippon's position in the world market.

An additional consideration that will affect Japanese steel production is the export of Japanese automobiles. In 1980, the Japanese automobile industry produced 11 million vehicles to establish itself as the world's leading producer. In that year, steel shipments to the automotive industry in Japan were 9.5 million tons. Automotive exports were 6 million vehicles, 2.4 million of which went to the United States. The automobile industry in the United States sought to restrict these imports and the effort resulted in a voluntary restraint placed on exports by the Japanese. How far this will be extended in the future is unknown. However, if the Japanese automobile industry meets with resistance to its exports not only in the United States but also in Western Europe it could reduce automobile production and thus curtail the demand for steel. In addition to automobiles, the Japanese export tractors, fork lifts, and a variety of mining, earth-moving, and other steel-containing heavy equipment.

The Japanese steel industry has concentrated very heavily in the past few years on exporting steel-mill equipment and making investments in foreign countries. Recently, Nippon Steel has signed a contract to construct a 600,000 ton per annum integrated steel works, based on direct reduction, in Malaysia. To achieve this, a consortium of Japanese companies was formed. Further, a joint corporation will be established between the Japanese and the Malaysians with the Japanese investing 30 percent of the equity. Nippon Steel's share of this will be 50 percent.

Other overseas investments by the Japanese steel industry include millions of dollars in raw materials, such as coal and iron-ore mines, integrated steel works, and oil and gas ventures. All five of the major integrated steel companies in Japan are involved in overseas investments. These include substantial sums in projects in the Philippines, Bahamas, Australia, the Middle East, Latin America, and North America.

Western European Exports

The EEC segment of the Western European steel industry will most probably have less than 160 million tons of reliable raw-steel capacity by the middle of this decade. If the planned modernization materializes and, since much of it is subsidized by governments, it undoubtedly will, many of the companies, although smaller, will be more efficient than they were in the late 1970s and early 1980s. Further, in a number of instances where the nationalized and subsidized companies have been relieved of at least part of their debt, the interest burden will be reduced and in some cases lifted, so lower total costs will result. This will enable the Western Europeans to compete more vigorously in the export market. They will meet the Japanese in the Middle East, the USSR, Africa, South America, and the United States, and in any year with less than strong demand, competition will be severe. One factor that could temper such competition would be a change in the relationship between supply and demand. This could be brought about by an increase in demand during the next few years or a reduction in supply. This reduction is taking place in the industrialized world and could have an effect on limiting destructive competition.

Current projections with respect to future demand in Western Europe indicate that even with reduced steelmaking capacity there will be considerable tonnage of steel available for export. In the short run, the 1982–1983 period, since the U.S. market is at least temporarily restricted by virtue of the countervailing duties and antidumping suits that are being pressed against Western European producers, a number of these producers will be impelled to find other outlets for their steel.

The United States as an Import Market

The United States will continue to draw heavy tonnages of steel imports through the remainder of the 1980s. With the reduction in U.S. steelmaking capacity to approximately 127 million to 129 million tons through the mid 1980s, the shipment capability of the industry, even with additional continuous casters, will probably be no more than 95 million tons and it could be somewhat less. In a year of reasonably good economic activity, it will be necessary to import 10 million to 12 million tons of steel products to supplement this supply.

This could be altered by the future development of the automobile industry. If the automobile industry does not recover to a point where at least 11 million vehicles are produced annually, this much additional steel will not be required. Nevertheless, a certain amount will be needed and it is reasonable to assume that the Japanese, the Western Europeans, and the Third World will be competing for this tonnage. If the Japanese continue to export large numbers of automobiles as well as other products made of steel, such as machinery and fabricated structures, to the United States, a significant tonnage of Japanese steel will be coming into the United States in addition to direct steel-industry products. Automobiles from Japan represent an indirect export to the United States of over 2 million tons of steel.

Attempts made in the United States to restrict imports over the last five years have been relatively unsuccessful. The trigger-price mechanism, installed in 1978, had some impact but has ultimately been judged a failure, with the result that antidumping suits and countervailing-duty suits have been filed in large numbers. The preliminary findings of the U.S. Department of Commerce relative to these suits indicated that substantial penalties would have to be paid on the imports in question, which would have acted to reduce the import tonnage.

On August 25, 1982, the Commerce Department released its final determination, and in nearly all cases, the duties assessed were substantially lower than under its preliminary findings. In the case of British Steel, the assessment fell from 40.4 percent to 20.3 percent. For Cockerill Sambre of Belgium, it was cut from 21 percent to 13 percent. In France, Sacilor's assessment was reduced from 30 percent to 20 percent; and Usinor's, on some products, from 20 percent to 11 percent, although in one instance it fell to only 18 percent. South Africa's duty was reduced from 12–16 percent to a de minimis level, as was that for West Germany, Luxembourg, and the Netherlands. Italy's assessment, however, was increased in one instance from 18 percent to 26 percent.

The Commerce Department stated that the final result was arrived at from an analysis of additional information and that it covered only 1.8 million net tons of imports in place of 4 million net tons. This procedure, which permits such severe changes in so short a period with the accompanying uncertainty to all concerned, demonstrates very clearly the need for an ongoing, dependable arrangement to govern steel imports into the United States.

Third World countries are anxious to participate in the U.S. market, particularly since they have built up new capacity which, in a number of instances, is in excess of their steel-consuming ability. Consequently, steel is available for export, and many of these countries hope that it will bring much-wanted foreign exchange.

Other Third World Export Activity

As previously indicated, the Third World is expanding its productive capacity and will become an increasing factor in the export market, vying for the traditional markets that the industrialized countries have supplied. Third World raw-steel production in 1980 was slightly in excess of 100 million tons. Plans for the next five years could bring this figure to as much as 125 million tons and a significant portion (probably half) of the increase will be marked for exports. This should come as no surprise since these countries are investing large sums in steel-production facilities and will attempt to export as much steel as possible to keep the new capacity in operation.

The exception to this trend is China, the largest steel producer in the Third World. It will probably not become a factor in the export market for a number of years since so much steel is desperately needed at home that imports will supplement domestic steel for much of the 1980s.

India, the third-largest producer in the Third World, has exported some tonnages of steel during the past decade with the figure reaching a high of 739,000 tons in 1975, after which it dropped off to virtually nothing in 1979 and 1980. India could be a factor in the export market from time to time; however, it is not expected to be a steady supplier of significant tonnages.

Some Third World countries have advantages in competing in the export market. First, their mills, as in the cases of South Korea, Brazil, Mexico, Taiwan, and, to some extent, India, are quite modern and capable of producing a quality product. Second, their wage rates are significantly below those of the industrialized countries. In the cases of South Korea and Taiwan, wages are about one quarter of those paid in Japan. And, third, most of the mills in the Third World are state owned and, if necessary, can be subsidized. In the next few years, an increasing tonnage of exports can be expected from at least three or four Third World steel producers.

International Steel Committee

In 1978, with many problems facing the world steel industry in terms of overcapacity, pricing, production, and destructive competition, the OECD met in October to organize an International Steel Committee with most OECD countries as members. The resolution that launched the committee stated the problems to be solved, the objectives of the committee, as well as functions that it was to perform.[5]

The problems characterized as widespread were

1. Persistent excess capacity;
2. Unjustifiably low prices on world markets;
3. Marked changes in traditional trade patterns;
4. Major dislocations of labor, frequently in high-unemployment areas;
5. Depressed financial performance among steel producers, restricting investment for modernization and rationalization of plants; and
6. Increasing governmental intervention in steel supply and demand, especially with foreign trade.

To address these problems, a series of objectives were drawn up, which proposed to:

1. Ensure that trade in steel will be as unrestricted and free of distortion as possible;
2. Encourage reduction of barriers to trade;
3. Enable governments to act promptly in crisis situations, consulting with interested trading partners;
4. Facilitate needed structural adaptations to diminish pressures for trade actions and promote rational allocation of productive resources to achieve fully competitive enterprise;
5. Ensure that measures affecting the steel industry are consistent, when possible, with general economic policies, taking into account implications for related industries, including steel-consuming industries;
6. Avoid encouraging economically unjustified investments while recognizing legitimate development needs; and
7. Facilitate multilateral cooperation consistent with the need to maintain competition in order to anticipate and to possibly prevent problems.

Membership in the steel Committee was open to all OECD countries, and several non-OECD countries with either substantial steel production or trade interests were invited to join.

One of the principal initial commitments made by the members of the committee was that "the members of the Steel Committee agree that they will avoid any form of competitive subsidization of exports or new steel-making plants and equipment and that the policies of government export-credit institutions regarding steel plants should be fully consistent with the international consensus developed in the OECD."

A second major commitment required that "the member governments agree that domestic policies to sustain steel firms during crises periods should not shift the burden of achievement to other countries and thus increase a likelihood of restrictive trade actions by other countries (for example, by artificially stimulating exports, by artificially raising stock levels, or by displacing imports)."

Further, as a general rule, domestic measures should not prevent marginal facilities from closing in those instances where facilities cannot become commercially viable within a reasonable period of time.

The two commitments, if followed literally, would have helped to solve many of the problems facing the steel industry in the 1980s. It is all too evident that there were a number of exceptions that have significantly detracted from the objectives of the International Steel Committee.

Four Third World countries—Mexico, Brazil, South Korea, and India—were invited to join the committee. These countries were very much concerned with the committee's commitment that involved restricting export credits to construct steel plants when they were not fully consistent with the international consensus developed by the OECD. All of these countries had substantial expansion plans in operation and did not wish to have them aborted. It is quite possible that South Korea would have joined the committee provided the other three countries did likewise; however, the invitation was rejected by India and Brazil and, consequently, South Korea. Mexico also declined, but keeps a special contact with the committee.

If the commitments and recommendations of the Steel Committee have been followed, some steel-industry expansion could have been curtailed and the closure of a number of marginal facilities accelerated, which would have aided considerably in reducing what the draft resolution called persistent excess capacity. However, national needs, problems, and crises took precedence over actions recommended by the International Steel Committee. A number of marginal facilities that could not be commercially viable should have been closed, yet they are still in operation since it has become politically difficult to terminate production at these locations.

The International Steel Committee has served as a meeting ground for the frank and open discussion of international steel problems and, in this respect, it performs a very necessary and beneficial function. Perhaps the original objectives were too lofty and, given the political realities, very difficult to achieve; yet it was a step in the right direction.

The steel industry's condition on a worldwide basis has deteriorated since the beginning of the steel depression in 1975. There has been some improvement in some years in a number of areas throughout the world; however, there has not been a universal recovery and, as of 1981 and early 1982, the situation has worsened dramatically. Trade restrictions, which the resolution founding the International Steel Committee abhorred, have multiplied and, short of a rapid recovery in steel demand throughout the world, will continue in operation. If one is to judge by present trends, they will probably increase and intensify. Virtually every steel producer in the world has restrictions on imports and most of the larger producers are actively promoting exports. There have been some suggestions that there should be an orderly marketing arrangement made for the world; however, these have been met with considerable opposition and it is doubtful that such an arrangement could be made that would be workable.

An example of how difficult it is to get agreement to foster orderly marketing was the particularly large tonnage of imports that came into the United States in January of 1982. There was some speculation concerning the possibility of limiting EEC exports to the United States for the first quarter of 1982 to 600,000 tons. In January 1982, however, exports from the EEC to the United States were 677,000 tons.

Trade in steel is necessary; however, the zeal to promote exports on the part of both industrialized and Third World countries has resulted in competition that can only be termed destructive in many areas in the world. Individual privately owned companies find it more and more difficult to remain in business under these circumstances without some government subsidy.

Approximately 60 percent of the world steel industry is government owned or heavily subsidized and this percentage will increase in the years immediately ahead to at least 70 percent, if not more. Currently, the areas that are unsubsidized include Japan, the United States, Australia, and a large portion of the Canadian companies; and there are also a number of individual companies throughout the world that are not government owned or subsidized. For example, 35 percent of the Brazilian production is in the private sector, as well as certain companies in India, Mexico, and Argentina. Virtually all of the European companies receive some form of subsidy either for research and development, capital investment, or the so-called emergency subsidies, which often provide funds to meet operating expenses. The struggle between private- and public-sector companies is at the heart of the international-trade problem in steel, and additional subsidies will only serve to intensify future trade disputes.

Notes

1. *Annual Statistical Report* American Iron and Steel Institute, 1969, pp. 32, 35.
2. U.S. Steel Industry Task Force, "A Comprehensive Program for the Steel Industry" (Washington, D.C.: Report to the President, December 13, 1977), p. 13.
3. Ibid.

4. *Monthly Report of the Iron and Steel Statistics* (Japan Iron and Steel Federation, July 1981), pp. 14–17.

5. Material concerning the International Steel Committee was drawn from an unclassified incoming telegram to the Department of State in October 1978, giving the text of the OECD Council's decision establishing the Steel Committee that was made public on October 27, 1978.

10 A Case of Survival

It is virtually certain that the world steel industry will not survive the balance of the 1980s in its present structure. The impact of recent adverse developments affecting the industry's basic condition and the strategies evolving—by choice or by necessity—in response to these developments make change, even radical change, inevitable on a national, regional, and global basis. Steel companies throughout the world will continue to face a variety of problems, and while some of these will be solved with difficulty, others may not be solved at all. By the mid-1980s, a number of companies will have reduced their capacities by closing selected facilities or will have lost their present identities through sale, merger, bankruptcy, or nationalization. There will also be significant and continuing developments in the application of new technology and the use of raw materials, while the most troublesome area in which problems must be faced and solved is, and will continue to be international trade.

In considering the prospects for the world steel industry in the decade ahead, the expectation is that changes in its structure will result not only from the corporate reorganization of individual steel companies, but also from (1) the planned or forced rationalization of their operations, (2) modernization and replacement programs implementing new technology, (3) continuing efforts to expand the production capabilities of developing countries, (4) government-subsidy programs, and (5) shifts in ownership between the private and public sectors. In the face of these changes, it remains to be seen whether or not the free-world industry will be able to survive as a profitable producer of quality steel products at competitive prices, or whether it will limp along as an ailing, predominantly government-owned and subsidy-ridden, inefficient entity that is destined to become immersed in recurring international-trade disputes, largely occasioned and aggravated by the differing objectives of its private- and public-sector companies.

Future Growth Limited to the Third World

The steel industry will witness very little growth through the mid-1980s. As of 1982, there are approximately 850 million tons of reliable steelmaking

219

capacity in the world and, by 1985, there will be at most 870 million tons. From 1986 to 1990, there will be some growth; however, very few plans indicating significant tonnages have been announced. Most projected figures for capacity in individual countries by 1990 are stated in very general terms, without reference to specific companies or facilities.

It is not surprising that the industry will witness very little growth in the years ahead, for growth depends on present demand and the confidence that it will increase in the future. The world steel industry is currently experiencing one of the lowest demand rates in the postwar period and most producers have very little confidence in the future. Steel companies in many areas of the world are in a state of confusion and, in some instances, almost bordering on despair. This is particularly true in Western Europe, in some parts of the Third World and, to some extent, in the United States. Without question this confusion has been intensified by recent decisions to impose countervailing duties to offset subsidies on imports to the United States. The decisions have had a sharp impact on the countries and companies involved.

In some parts of the world, 1982 will mark the postwar low in the rate of steel industry operations and financial returns. The United States has been particularly hard hit, as its production for the first half of 1982 dropped some 35 percent below that of 1981. Further, the U.S. operating rate declined to 43 percent of capacity, a figure not seen since the depression of the 1930s. Canada also witnessed a sharp decline of 20 percent below the first half of 1981. By contrast, during the first six months of 1982, the steel industry of Japan actually increased its production by 4 percent above the same period in 1981. The EEC showed an extremely minor decline of 0.4 percent for the six-months' period. However, projections for the second half of 1982 for both Japan and Western Europe indicate a considerable decline. The Japanese are projecting a drop in production which could bring them below the 100-million ton level for the year. Further, evidence of the decline expected in the EEC was the restriction placed on steel production for the third quarter by the EEC Commission, which will cut raw-steel output some 4 million tons below the third quarter of 1981.

Brazil, the largest steel producer in Latin America, experienced a 2-million-ton drop in production in 1981, its first decline in a decade. In the first half of 1982, production was down approximately 14 percent, indicating a further significant decline.

Large financial losses have been recorded for the first half of 1982 by many steel companies throughout the world, and these losses are expected to worsen by the end of the year. Consequently, it is necessary to look beyond 1982 for a revival in the world steel industry.

The steel recession and, in some areas, depression of the late 1970s and early 1980s brought with it a basic change in philosophy among many steel producers. They no longer think in terms of producing large tonnages, but rather in terms of profitable tonnages and, with that in mind, many have eliminated marginal and unprofitable facilities as well as some products. This is particularly true in the United States where, in addition to eliminating facilities and products, steel companies have diversified into other areas to achieve improved profitability. In recent years, the Japanese also have concentrated on more profitable products, abandoning the tonnage approach and, to some extent, their coveted goal of high market penetration.

In keeping with this philosophy, integrated steel companies in the United States closed 12.5 million tons of raw steelmaking capacity between 1977 and 1981, while British Steel Corporation cut its capacity in half. By 1985, U.S. producers will close an additional 8 million to 10 million tons, leaving a reliable raw steelmaking capacity of 127 million to 129 million tons. It is interesting to note that the integrated producers in the United States have also relinquished a number of products to the smaller electric-furnace mills.

In the years ahead, billions of dollars will be spent throughout the world to improve facilities, operations, and product quality by the installation of more continuous casting, as well as new rolling and finishing mills, such as the seamless pipe and tubing mills now under construction. In spite of these expenditures in steel, a number of companies will become much more diversified, as they invest in other, more growth-oriented activites to improve their profit positions.

In Western Europe, a declining market and drastic financial losses have curtailed any thought of expansion. Rather, the industry has contracted and will continue to do so. Steelmakers in the EEC have more than adequate capacity to meet their current needs and are extremely pessimistic about the future, an attitude that is having a major impact on their management decisions. Current reliable steelmaking capacity in the EEC is approximately 160 million tons, and this will be further reduced by a minimum of 5 million tons or, more likely, 8 million tons in the years ahead.

The EEC's reliable capacity of 160 million tons differs from some published capacity estimates approximating 200 million tons. A breakdown of these estimates shows much larger capacities for West Germany, the United Kingdom, France, Belgium, and Italy than are justified on the basis of a plant-by-plant analysis. For example, the figure for West Germany is put at approximately 68 million tons, while realistically it is at best 55 million tons. In fact, West German steelmakers acknowledge that if a boom occurred within the next year their industry could produce no more than 50 million tons.

The EEC has plans to improve its finishing facilities and increase continuous casting while reducing its raw-steel capacity, so that the surviving units will be much more efficient. Some of the capital to install these improvements will be made available by government grants, so that it is quite likely the planned improvements will be carried out.

During the 1960s and much of the 1970s, the Japanese built a highly efficient steel industry, which has dominated the export market for more than a decade. Its current capacity of 145 million tons will probably be reduced by a few million tons in the years ahead and, in spite of the fact that their existing facilities are relatively new, the Japanese will invest large sums to improve efficiency and develop quality products. This program has been inspired in part by new capacity in the Third World that will diminish the Japanese export market in certain areas, so that the Japanese must be prepared to attain profitability at lower operating rates.

The other industrialized countries, such as Canada, Australia, and South Africa, will add very little to their steelmaking capacity within the next five years. Canada has expanded within the last three years and must adjust to this new steelmaking potential, while the size of South Africa's industry will remain static, and Australia will reduce its raw steelmaking capacity and improve its finishing facilities.

As previously indicated, the Third World has plans—in some countries ambitious plans—for expansion, but many of these, announced as recently as 1979 and 1980, already have been materially reduced. Those that materialize, for the most part, will be carried out with government guarantees of loans or direct government investments.

Since most of these programs will provide steel, not only for domestic use but for exports, it is incumbent on these countries to examine the world steel industry in terms of its supply capabilities. For example, the Third World countries, particularly in the Far East, have installed and plan to install several wide hot-strip mills. One is under construction in Taiwan; one has just been completed in South Korea and another is planned; and still another is under construction in Indonesia. There is an excess of wide hot-strip mill capacity on a world-wide basis. Europe, Japan, and North America have more than enough capacity to take care of the entire world demand for wide sheets, much of which was built in the 1960s and 1970s, and a significant portion of which is now idle with little hope of recovering to a point of full utilization. The addition of more wide strip mills to provide steel sheets for the world market could lead to severe price competition, which would result in significant losses.

In assessing the possibilities for growth in the years immediately ahead, it must be kept in mind that the steel industry is now suffering from over-capacity, which has brought on intensive and sometimes destructive international competition. On a global basis, no more capacity is needed, and a worldwide moratorium on expansion would be most welcome to ensure future profitable operations. This moratorium can be expected in the industrialized countries, due to the present depressed and unprofitable state of their steel industries. Some Third World countries, however, feeling the need to produce more steel to reduce their dependence on imports, still plan to extend their capacities. Whether this can be achieved in the years ahead and in the proportions desired remains to be seen. Worldwide economic conditions in 1982 indicate that such installations will be stretched out over a longer period of time, if not postponed or cancelled, as in the case of China and South Korea. On balance, however, Third World capacity will continue to grow.

Raw Materials Remain Abundant

An abundance of iron ore exists throughout the world that will care for the needs of the steel industry indefinitely into the future. Until 1985, there will be very little need to develop additional ore deposits to supply the world's steel requirements. However, new mines that involve enormous expenditures of capital may well be needed toward the end of the decade, as the presently mined deposits become depleted and steel production grows.

The 1982 depression in steel has resulted in the temporary closure of many ore mines and beneficiating facilities, and it will take time and a significant increase in demand for the iron-ore industry to resume operations at close to full potential. There is very little reason, therefore, to look for new ore developments before 1985. The opening of an ore mine, however, is not an overnight proposition, it requires a number of years and a large investment, particularly if the ore is not advantageously located. This is the case with respect to the Carajas ore deposit in Brazil, probably the largest new development in iron ore. In addition to opening the mine, the location of the deposit requires the construction of a 1,000 km railroad over difficult terrain, as well as the development of a town at the mine site and a port on the ocean through which the ore will be shipped.

The production of Carajas will not be needed until the late 1980s, but the mine must be developed prior to that time if it is to operate when its output is required. Thus, in the years immediately ahead, there will be investments in new iron ore deposits, although those currently in operation are more than adequate to service the worldwide steel industry through the mid-1980s and somewhat beyond.

Financing for iron-ore projects is more difficult to obtain when the world steel industry is in a depressed state, even though it is appreciated that they will be needed a few years hence. Because of this situation, it is necessary to have some type of government financing, since private companies at the present time either do not have or do not wish to risk the capital. For example, the Carajas project is dependent to a great extent on the World

Bank, the Export-Import Bank of Japan, and the EEC's financing, as well as that of the Brazilian government. In addition, a group of Japanese steel-makers have agreed to provide a loan. On a worldwide basis, it is highly unlikely that any iron-ore deposits will be developed entirely by private capital in the next five years.

The currently operating iron-ore mines in Brazil, Australia, Venezuela, South Africa, and Canada have accounted for a major share of the world's iron-ore production and trade and will continue to do so. These operations represent large investments, the recovery of which will foster continuing competition for the available world market.

High-quality metallurgical coal is available, but it is by no means as abundant as iron ore. The United States, Poland, and Australia will continue to export large tonnages, with lesser amounts coming from Canada and West Germany. Efforts will be made to reduce dependence of the coking process on high-grade metallurgical coal by the substitution of lower-grade coals and various improvements in technology, such as the briquetting of coal. Formed coke will not be available in any significant quantities before 1990.

Because of wide variations in the quality of coking coal and difficulties in mining it, its production is subsidized in some countries, such as the United Kingdom and West Germany. There is little hope that these subsidies will be discontinued for some time, since they are heavily involved with employment in both countries, which makes this a political as well as an economic issue.

Over the remainder of the decade, ferrous scrap will be the overwhelmingly dominant feed for electric furnaces, despite the projected increase in the production of direct-reduced iron (DRI). The availability of quality scrap, however, could and probably will act as a restraining factor on the worldwide growth of the electric furnace, particularly as a producer of quality carbon steels.

New Technology Essential but Costly

Much of the world steel industry employs the latest technology. The more fundamental breakthroughs, such as the basic-oxygen steelmaking process and continuous casting, are used throughout the world. There is scarcely an integrated steel mill, outside of the Soviet bloc, that does not employ the basic-oxygen process in one form or another. Up to the mid-1980s, this process will grow more as a replacement for the open hearth than as an expansion of steelmaking capacity.

Electric-furnace innovations will continue on a broad front, encompassing additional advances in ultrahigh power, oxygen injection, continuous charging, raw-materials preparation, water cooling, and ladle refining. Water-cooled sidewall and roof panels will come into widespread use; additional quantities of scrap will be fragmentized and preheated for through-the-roof charging; direct-reduced iron will come into wider use as a scrap supplement; and further improvements in electrode performance will derive from changes in design and the use of coating and water-cooling. All of these innovations will contribute to the effective, economical use of the electric furnace.

Continuous casting has been widely adopted throughout the world, being used to process 33 percent of global steel production in 1981. It will increase in its penetration in the years ahead; however, the limiting factor will be the availability of capital to invest in this facility. Major units, such as those being installed in industrialized countries, that are capable of a million tons or more of production, are expensive, costing as much as $100 million. In spite of this, top priority has been placed on the installation of continuous casting in a number of nations, particularly in the United States and Western Europe. The opportunity for extracting further benefits from the process is not as great in Japan, which in 1981 already had 70 percent of its steel continuously cast.

On a worldwide basis the amount of continuous casting could increase to more than 40 percent of the steel made by 1985. This will permit, because of the higher yield obtained, more finished products to be shipped from the crude steel produced at that time.

In the years ahead, increasing attention will be given to new iron- and steelmaking processes having major potential to reduce both capital and operating costs. The world's direct-reduction capacity will increase significantly by mid-decade, reflecting in part the ongoing trend toward steelmaking in modern electric furnaces equipped to derive major benefit from the use of DRI. The major concentration of new direct-reduction activity will be in Third-World countries with ample and economic natural-gas supplies, while the industrialized world will, for the most part, await further advances in direct-reduction techniques based on coal and will obtain relatively small but increasing quantities of DRI from overseas producers with outputs exceeding their domestic requiements.

While the production and use of DRI continues to grow, additional attention will be given to new and emerging technologies for the production of molten pig iron without metallurgical coke or electric power and for oxygen steelmaking from solid charge materials. The commercialization of such radical approaches to iron and steelmaking eventually could play a major role in assuring the survival of integrated steel producers currently facing a variety of production, environmental, and capital-cost problems in the hot-metal end of their operations.

The major problem in terms of the world steel industry's utilization of new technology is not the availability of innovation but the capital costs involved. However, because advances in technology are essential to the industry's survival, joint ventures can be anticipated not only within countries but on an international scale. Recently there have been suggestions of joint ventures between Japanese and American steel companies, as well as the Japanese and Western Europeans, and although the results thus far have been limited to know-how agreements it now appears that Nippon Kokan will invest in the Rouge Steel Company in Detroit.

Although the adoption of new or improved technology is deemed necessary for the survival of a steel company, the capital costs involved are very often so great that they neutralize much of the potential economic benefit. For example, a modern continuous wide hot-strip mill with an annual capacity of 4 million tons involves an investment of some $400 million. If this is depreciated over a five-year period, it would involve a charge of $80 million per year, or $20 per ton. Further, if one assumes that half of the investment is borrowed at an interest rate of 15 percent, this would involve $30 million per year, or $7.50 per ton. It is understood that interest will decline as the loan is amortized; however, if its amortization is over a twenty-year period, this would involve $10 million per year, or some $2.50 per ton. Thus, the total capital charges on the mill would be about $30 per ton. The advantages to be obtained from the installation of the new mill would have to be considerably more than $30 per ton to make the project worthwhile.

The same is true of other large facilities, such as blast furnaces and coke ovens. One company contemplating the installation of a blast furnace to replace 2 million tons of existing capacity found that it would involve an investment of $300 million. If written off in five years, this would represent $60 million per year, or $30 per ton, for depreciation alone. Even if the facility were built with the company's own funds and no money were borrowed, the $30-per-ton figure and the large capital investment were enough to force a decision by management to indefinitely postpone the project. Of course, capital charges can be spread over a longer span by lengthening the depreciation period, but this leaves the company vulnerable to inflation.

With the high capital costs involved for the installation of modern technology, many steel producers have had second thoughts and are examining carefully the possibility of upgrading existing facilities rather than replacing them with new units. In the years ahead there will be some of both. A new unit is needed if the existing facility cannot make the quality necessary to compete. If, on the other hand, an upgraded facility can make a quality product that is competitive, then the new unit will most probably not be installed. Such a situation has developed worldwide with regard to seamless pipe facilities. Some existing units can be upgraded to meet the high-quality standards demanded by oil-well drillers today, while others cannot be improved no matter how much is invested. Thus, new facilities will have to be installed, if the pipe producer wishes to remain competitive.

In the installation of new or improved technology, cost differentials between countries, as well as the type of ownership, of a company, play a significant role. Installation costs in Japan and some Third World countries are decidedly lower than in many other areas throughout the world, and a government-owned company has an advantage over a privately owned company, since the former can obtain capital to install new equipment from the public treasury while the latter does not have this luxury and must raise its own funds.

This contrast in capital investment between publicly owned and privately owned companies is pointed up by recent decisions made on the replacement and modernization of plant and equipment. In the United States virtually all of the major steel producers have cut back or stretched out their capital investment programs. Because of the current steel depression and heavy losses incurred in connection with steel operations, United States Steel, Bethlehem, Armco, and other producers have announced that a number of capital projects will be deferred. In contrast several Western European countries which have incurred losses are proceeding with capital investment because of government grants made available for this purpose. Likewise, Sydney Steel in Nova Scotia, which has been in financial difficulties, has been granted $96 million by its government and is proceeding with a replacement program.

Steel companies, because they must compete not only among themselves on a domestic as well as an international basis but also with companies producing substitute materials, must if at all possible avail themselves of new technological developments, particularly if these are radical enough to reduce costs and improve quality substantially.

Shift to Public Ownership Continues

The ownership of the worldwide steel industry has shifted dramatically since 1950, when it was predominantly in the private sector. As previously indicated, some 77 percent of the steel produced in 1950 came from companies that were privately held. In 1981 approximately 60 percent of world steel production came from companies that were either government owned of heavily subsidized.

Much of this shift has been due to the growth of steel output in the USSR from 27 million tons in 1950 to 151 million tons in 1978, its peak production year. Likewise, steel output in the Eastern European Communist countries rose from 8 million tons in 1950 to over 60 million in 1980. In addition to these developments, there have been significant ownership changes in Western Europe as many private companies have either been nationalized or forced to apply for large government subsidies in order to survive. This trend toward nationalization or government subsidies has had a definite impact on steel-industry operations.

The prime and demonstrated concern of government-owned companies has been to maintain employment, a concern that is economically as well as politically and socially motivated, especially during times of recession or depression. This has been experienced by the steel industry in many parts of the world, almost without interruption, since 1975.

The attempt to maintain employment forces the companies under government pressure to continue operations, and in distressed times their steel production usually exceeds demand. Consequently, it must be sold wherever possible at low prices, which have a disrupting effect on the domestic as well as the international steel market.

This places the privately owned companies at a distinct disadvantage, since they have no outside means of absorbing the losses incurred in cutting prices to meet the subsidized competition. Government ownership or government subsidies, when they result in price cutting to preserve employment, are deplored by all, but practiced by many. Almost without exception, when steel companies are nationalized, the government in question as well as the new management stoutly proclaim that the newly organized company will be operated as if it were privately owned and will be profit oriented. This statement was made, unquestionably with good intentions, when British Steel Corporation was formed in 1967, but when the industry came on hard times and losses appeared those losses were absorbed by the public treasury so operations could continue and layoffs were prevented.

In favor of this position, it is urged that it is better to have people employed by a company that is losing money and subsidized by the government than for these people to be unemployed and on the dole. Undoubtedly, working is better for individual morale than receiving relief. However, when the products that result come into competition with those of private companies struggling to remain profitable, the situation is disruptive and leads to heavy losses, as happened in the Common Market during the late 1970s. The inevitable consequence is and has been that more companies faced with financial collapse, look to the government for assistance, and this extends government ownership and influence further into what was once the private sector.

At the present time, particularly in Western Europe and the Third World, the tendency toward government ownership and subsidies is growing. Most companies in Europe of any size receive some type of government subsidy. These include subsidies to meet operating expenses, grants for research and development, and subsidies for capital investment. The subsidies for capital investment are either outright grants, representing a portion of the total investment, or the absorption of part of the interest paid on loans incurred for capital investments. In the Third World, a great part of the steelmaking capacity installed in the past five years is government owned.

In the coming decade, particularly during depressed times, government ownership or subsidy will extend further into the steel industry. In many instances, this will be done to equalize the advantage of a government-owned or subsidized company. An example of this is occurring in West Germany, where the reorganization of part of the steel industry is planned for the years immediately ahead and will be substantially helped and, indeed, made possible by government grants.

With the rising cost of new facilities, companies that can call upon the public treasury for financial assistance have a distinct advantage over their private counterparts in upgrading their plants. As a defense against this inequitable situation, many private companies have no choice but to shrink in size, modernize as fully as possible, and diversify their operations into nonsteel areas in order to maintain some degree of profitability. This has already taken place both in the United States and Western Europe.

The United States Steel Corporation, the largest in America and second largest in the Free World, has reduced raw-steel capacity, modernized or is modernizing its remaining steel facilities, and has made a huge investment in the oil industry. This was done to ensure future profitability since its steel operations were unprofitable in a number of the past years.

Another company, National Steel, one of the seven largest producers in the United States, has reduced capacity in line with realistic expectations of market demand. It has also improved steel-making facilities by adding continuous casting and has diversified into aluminum and financial services.

In Western Europe, Thyssen, a private steel company and the largest in West Germany, is a highly diversified enterprise that has reduced steel capacity and installed new facilities.

Since shrinkage, modernization and diversification are and will continue to be the defenses of privately owned steel companies in their competitive struggle with government-owned and subsidized companies, the private sector of the steel industry on a worldwide basis will account for an ever-diminishing percentage of total steel production. This is particularly true, since the new capacity coming onstream in the Third World, in such countries as China, Mexico, India, Brazil, South Korea, and Taiwan, will be overwhelmingly government owned or subsidized.

International Trade Problems to Recur

The growth in international trade has made steel available throughout the world and, at the same time, has generated a number of pressing problems. The ninefold increase in steel traded internationally—from 20 million tons of raw-steel equivalent in 1950 to 180 million tons in 1980—is the result of a fourfold increase during the same period in the worldwide growth of steel capacity and production. It is expected that trade in the years ahead will at least maintain the 1980 level since there are steel-deficit areas in the world and the steel industries of many producing countries have more capacity than is needed to satisfy their domestic demand. Therefore, they must export.

Trade between steel producers and countries with steel deficits usually does not present a problem. There are also no problems in periods of strong steel demand, such as in 1973 and 1974. During those years, steel consumers everywhere were so anxious to obtain enough to meet their requirements that they were willing to buy steel wherever they could at almost any price. All restrictions on trade were forgotten. The problem arises in periods of slack business when subsidized steel is sold below cost in foreign markets.

Crises have arisen from time to time in the late 1970s and early 1980s, which, unfortunately, have pitted the publicly owned and subsidized companies against their privately owned counterparts. Relief has been sought by restricting imports through a number of means, including the trigger-price mechanism in the United States, designed principally to force exporters to the United States to charge prices that would cover their costs of production; and in Western Europe, plans developed in the EEC requiring steel producers to maintain production quotas and price levels. These were temporary stopgaps, for all agree that the long-term solution that would solve the problem both in Western Europe and the United States is to end subsidies so that government-owned or subsidized companies would have to survive on their own and thus price products accordingly. Steps have been taken in the EEC to achieve this goal; however, it will be some years before the objective is realized. The Western European problem persists and will continue until subsidies have been withdrawn or, alternatively, and this is not an impossibility, until all companies are equally subsidized. If the latter move is taken, it would be unfortunate, since it could place the entire EEC steel industry under government domination.

Unless subsidies are terminated, restrictive devices will be maintained as a protection against them. This is unfortunate since protective measures are not a long-run solution to the trade problem. A free market with competitive forces at work would be the healthiest approach for the worldwide steel industry. However, as long as a number of companies, and there are many, continue to be supported and subsidized so they can cut prices drastically and still survive, the free market is in jeopardy.

It is regrettable, but subsidies and government ownership have become a way of life for the steel industry in many parts of the world. The situation could improve since subsidies for capital investment in the EEC should diminish and possibly disappear when the industry is rationalized and modernized. The date set for this is 1985 and hopefully it can be achieved.

Trade Areas

In world steel trade, the market in the United States has become the object of virtually every steel-producing country. It is the largest in the industrialized world and in many respects the most lucrative.

A number of countries, particularly Japan and those in Western Europe, have been exporting large tonnages of steel to the United States for more than fifteen years. In the past five years, annual imports from the EEC have been as high as 6 million tons and those from Japan were in the same range. This steady participation in the U.S. market has led the Japanese and the countries in Western Europe, because of long standing customer relationships, to talk of "their share" of the U.S. market. It is the only market in the world in which outsiders refer to as having a definite share. No outside country refers to its share of the Japanese market or its share of the EEC market.

In addition to the traditional exporters to the United States, some of the Third World countries that have recently built up significant steelmaking capacity have shipped sizeable tonnages into the U.S. market. This is particularly true of South Korea, which with its new-found steel capacity has shipped an average of 1 million tons per year into the U.S. market from 1978 through 1981. Brazil has also increased its exports to the United States. The Brazilians' desire to ship to the United States was evident in discussions held during the annual meeting of the Brazilian Iron and Steel Institute in April 1980. One of the principal questions raised was "How can we ship more steel to the United States?" Imports from other countries that have recently enlarged their steel industries, such as South Africa, have also increased appreciably during the last five years. In examining the growth of imports, they can be attributed to the desire of the steel producers throughout the world to share in the U.S. market, as well as to a very active group of importers in the United States.

In discussing the market penetration of imports into the United States, which in some years has been as high as 19 percent, it is often stated that the U.S. steel industry is at a disadvantage because it has not modernized its plant to compete with foreign producers. This statement has been repeated over the past twenty years. It had a great deal of truth in the early and mid-1960s, when imports came in from new plants abroad that produced at low costs due in great part to modern plants and low labor rates. In the

1960s and 1970s, the industry in the United States spent some $35 billion on plant and equipment and now is a modern industry in many respects although much remains to be done. It should be borne in mind, however, that even a fully modernized facility cannot compete with steel that is heavily subsidized and priced, because of that subsidy, 15 percent to 20 percent below its cost of production. For the remainder to the 1980s the United States will continue to be the largest net importer, not only of steel but also of products containing steel.

The EEC is another area in which there is a great deal of trade in steel. However, most of the trade is among the members of the Common Market and, in some years, a total of more than 30 million tons has been shipped across its international borders. In addition to intra-Common Market trade, a relatively small amount of imports comes from outside areas including Western Europe and Japan.

Although Japan is the world's leading exporter of steel, it has received very little in the way of imports over the past twenty years. In many years imports were as low as 200,000 tons, while exports were well in excess of 30 million tons. In the past two years there has been a change, as Japanese steel imports exceeded 1 million tons annually, with South Korea and Taiwan as the leading suppliers. The Japanese report that the import tonnages have included "considerable amounts of secondary grade and defective quality goods" and that the prices of these imports in Japan "appeared to be much lower than the domestic selling price by Japanese steel producers." Since F.O.B. prices in the exporting countries are presumed to be much lower still, the Japanese have concluded that "the recent dramatic increase of steel imports may be attributable to prices which are below cost or to government subsidies."[1] However, it is possible that some costs in South Korea and Taiwan are lower than in Japan, given labor rates per ton that are only one-third as great and the use of the latest production technologies that have been implemented, often with Japanese assistance, at large seacoast plants that number among the newest facilities in the world.

The Japanese have been very much concerned with this development, as can be seen from the fact that some statements claim Japan is being affected by Korean imports. On the other side, the Koreans state that their balance of trade with Japan is so negative that it permits this tonnage of exports to Japan.

There has been continued pressure on Japan from its trading partners to open its doors to imports from the rest of the world. Some concessions already have been made by the Japanese; however, it is doubtful that steel imports will amount to very much more than they are at present by the mid-1980s.

The Third World has been a market for steel products of the industrialized world during most of the post-World War II period. However, during the past five years, the installation of a substantial amount of steelmaking capacity in many areas throughout the Third World has reduced the steel market available to outsiders. This will continue in the years ahead, although capacity installation will be slowed considerably.

Some countries, such as Brazil, South Korea and Taiwan, have become net exporters of steel and thus will compete with the industrialized countries. However, there are a number of steel-deficit areas in the Third World that will continue to provide a market for exports.

Proposed Solutions

A number of solutions have been proposed to solve the trade problem as it affects various countries. These include orderly marketing agreements, both limited and general in scope, voluntary quotas, trade-restrictive legislation; subsidies for companies not currently subsidized; and a program to reduce steelmaking capacity, particularly in the EEC. The principal problem to be solved is access to the U.S. market for without this the EEC and Japanese steel industries would be severely damaged. On the other hand, the industry in the United States has resisted and will continue to resist what it terms excessive imports that have in recent months taken as much as 25 percent of the U.S. market.

Strenuous efforts have been made by steel companies and governments in both the EEC and the United States to reach an equitable arrangement. The Japanese, in late 1981 and throughout the first half of 1982, have tried to mediate this dispute; however, they are not a disinterested party, since any quota arrangement would have to include them. The major problem in international trade is the determination of the amount of steel that outside companies can send into the U.S. market. It is in the interest of the U.S. steel companies to minimize this amount, while the EEC, Japanese, Canadians, and the Third World are interested in obtaining as large a share as possible. Further, the U.S. situation is unique insofar as the U.S. government and the U.S. steel industry are being asked to bargain away a share of the market. No other steel industry in the world is in this position.

The recent decision of the Commerce Department indicating that several EEC steel companies have been subsidized to a large extent and that, therefore, countervailing duties must be imposed, has sent shock waves through the EEC. The reaction indicating the extent to which these countries depend on the U.S. market was summed up by the EEC Commission when it stated: "This decision will disrupt traditional trade flows, virtually eliminate exports of substantial value from certain member states and have serious indirect effects on other member states."[2]

The EEC responded by proposing various strategies to circumvent the

Commerce Department decision. These include the swapping of tonnage so that the West Germans and Dutch, who are not subject to countervailing duties, will take a larger share of the U.S. market and relinquish some of their other markets to those companies who are subject to countervailing duties.

The Western Europeans, including the West Germans, expressed further concern over the Commerce Department's decision, maintaining that the tonnage that would have been shipped to the United States will now be thrown on the European market and create havoc. This was expressed in a recent interview with one of the leading West German steel producers who said, "We have to compete against subsidized competitors in Europe and we get all the tonnages back which they can't export to the United States."[3]

Negotiations between the EEC and the United States are continuing in an attempt to reach some accord on the tonnage of steel that the United States can accept. As indicated previously, the United States needs steel imports; however, it does not need the amounts of imports that have come into its market in 1981 and during the first six months of 1982.

The British were particularly hard hit by the decision since the countervailing duty on their products amounted to 40 percent. This brought a comment from the British Trade Secretary that the Commerce Department's findings jeopardized BSC's program to improve its efficiency and break even on its operation within the next year.[4] The seriousness of these statements reinforces the judgment that the EEC steel industry feels it will have difficulty in surviving without access to what is considered its share of the U.S. market.

On the other hand, U.S. producers strongly resented the penetration of their market to the extent of 22 percent for the first five months on 1982 and have responded vigorously. David M. Roderick, Chairman of United States Steel Corporation and the American Iron and Steel Institute, said in relation to the import figure of 1.7 million net tons in May 1982, "Imports are on the rise again, reflecting the policy of governments abroad to maintain production in their mills at whatever cost, and regardless of the state of the world market. . . . Imports of this magnitude—representing an unacceptable 25 percent or more of the total apparent supply in the domestic market—are a damaging burden on the American steel industry and its employees. Each million tons of steel represents around 5,000 jobs per year; and to the extent that these imports reflect subsidization by foreign governments, or sales at less than fair value, the jobs of American steelworkers are being stolen. At a time when our industry is operating below half of capacity and a third of our employees are out of jobs, this is an unacceptable state of affairs."[5]

Both sides in the dispute have taken strong positions, which make a compromise more difficult. Some of the negotiations that took place after the Commerce Department's decision on June 12, 1982, indicate that the

EEC producers would settle for a 5.8 percent share of the U.S market, while the U.S. position is for less than 5 percent. There is also resentment among U.S. producers at the thought of giving away a percentage of the market, for not just the European share is at issue. The Japanese would want at least an equal share and others, such as Canada and some Third World countries, would also look for a share of the U.S. market. The total penetration of 22 percent in the first five months of 1982 is regarded by the Americans as completely unacceptable.

Temporary solutions in the form of quotas or orderly marketing arrangements will be offered and it is possible that some may be accepted. However, a permanent solution and, indeed, it seems the best solution to the trade problem would be the withdrawal of all subsidies, so that steel companies throughout the world would have to operate on a competitive basis. Unfortunately, this is not realistic, in view of the fact that too many companies are government owned and in times of distress, in order to maintain employment, would most probably be subsidized despite any intention to the contrary. High unemployment has a strong political and social impact, and remedies are considered a top priority. Thus, it seems that as long as there are government-owned and subsidized companies and steel is sold in the world market during depressed periods, prices will be cut to sustain operations to alleviate unemployment.

Other solutions that have been applied in limited areas include the imposition of production quotas and price levels. These have not been entirely successful but have helped to ease the problem. An outstanding example is the current program in operation in the EEC. The EEC has also been reasonably successful in controlling imports into its market by means of quotas and set prices that are arrived at through negotiations with individual countries outside of the EEC. Obviously such an arrangement would be difficult to extend to other areas since it would involve more countries, and authority to impose sanctions for violations would be extremely hard to obtain.

An attempt was made toward the solution of the world trade problem by the establishment of the International Steel Committee. This group serves as a very useful forum for the exchange of information and the expression of attitude.

Because the trade problem is fraught with so many difficulties, it appears that most countries will continue to protect their interests by restrictive trade measures. However, this is not the best solution since protective devices do not provide a permanent answer. Further, if they are too widespread there is a tendency to limit competition and the steel industry needs competition, not of a destructive nature but rather an amount that fosters new technological developments that help to reduce costs and improve product quality.

An example of how the lack of competition can stifle the introduction of new technology is the Soviet Union. Its steel industry does not operate in a competitive market, therefore, there are no pressing demands to improve technology. As a consequence, the Soviet Union in 1981 produced 60 percent of its steel by the open-hearth process, which the rest of the world regards as obsolete and uneconomical.

Access to the U.S. Market

Access to the U.S. market is the central problem in international trade in steel. This has been a recurring difficulty for the last fourteen years, and a number of attempts to solve it have had no lasting effect. Currently the U.S. market has no ongoing defense against unfair import competition. There are trade laws which require a domestic steel producer to prepare a complaint to prove that he has been injured by unfairly priced steel in the American market. However, this procedure is costly and injury often difficult to prove. In most cases; it takes a year or more for the steel producer to prepare a suitable complaint, setting forth a prima facie case of trade law violations, and after that it can take as much as thirteen months from the initial filing of the complaint before the finding is published and duties are assessed against the violator. The cost and time involved make it almost prohibitive for a small producer to prepare and process such a case. The time lag between what is conceived by the producer to be injury and the ultimate resolution of the case is so lengthy that irreparable damage could be sustained in the intervening period. Further, uncertainties arise during this extended period that threaten to damage valued international trade relationships between traditionally friendly trading partners. Thus there is no question but that an ongoing automatic solution to the problem is a necessity.

It seems evident from past experience that a solution to the problem must involve tonnage limitations. An amount stating the total tonnage of imports permitted into the United States should be established and this divided among all the countries that signify their intention of participating in the U.S. market. The amount can be an absolute tonnage figure adjustable annually or, what would be more difficult to determine and administer, a percentage of the U.S. market. This, for example, could range from 12 percent to 15 percent. The market could be estimated quarterly and corrections made on the same quarterly basis. The tonnages allotted to individual countries could be based on their historical participation in the U.S. steel market during the span of recent years.

For the arrangement to be effective:

1. All products should be included, with the possible exception of semi-finished forms restricted to slabs, billets, and blooms.

2. Geographic distribution within the United States should be taken into account.
3. The agreement should contain provisions to regulate product mix so that the market for an individual product could not be inundated.
4. The tonnages and products involved should be strictly monitored by the exporting countries, as well as by the United States.
5. The arrangement, once concluded, should have some permanence, at least five years and possibly longer.

Past failures of mechanisms dealing with prices require a shift to a tonnage basis. Trigger prices were circumvented, challenged, and ignored. They were circumvented in a number of ways, including the establishment of outside corporations offshore to which steel was sold at less than the trigger price and then resold to a company in the United States at trigger prices. Another means, employed by the U.S. unit of Mitsui, a Japanese trading company, over an extended period, involved submitting fictitiously high prices to the U.S. Customs Service, while actually selling steel to the customers at lower prices. Trigger prices were challenged as too high by the Western European producers when currency exchange rates allowed them to produce steel at a lower dollar price, which they claimed was beneath the trigger price established on the basis of Japanese costs. Trigger prices were ignored when a number of companies, having been denied preclearance, chose to ship steel into the United States at prices considerably below the trigger and risk an investigation.

Further, since government ownership and subsidies for the steel industry will continue in many countries, there will always be a question of subsidized steel prices. This, coupled with the trade laws on the books in the United States, will bring continued suits, making for uncertainty in the market.

The world steel industry, outside of the United States, particularly that of the EEC, is anxious to negotiate an agreement on access to the U.S. market. This is reminiscent of events in the summer of 1968 and the fall of 1977. In both instances, the Western Europeans and the Japanese proposed voluntary restraints on their U.S.-bound exports under the condition that more severe restrictions, then pending, be withdrawn. In 1968 Congressional hearings for proposed legislative quotas were involved; in 1977 the penalties assessed in the Gilmore antidumping case were involved, as well as the desire to head off potentially adverse decisions in a number of other antidumping complaints.

The present offer to negotiate involves a provision that the Commerce Department's judgments of the Trade Law violations be set aside. The proposal for a limitation on imports under this offer seems somewhat similar to that made fourteen years ago; however, the world's steel problems are much more serious now than they were then, and negotiated trade agreements are much more difficult to structure and effectually administer now that the world steelmaking capacity is dispersed among a greater number of steel-producing nations. Nevertheless, a solution must be found. Further, since it is the U.S. market that is being sought, and the United States has the second-largest steel industry in the world, the U.S. industry feels that it behooves those who export steel to the United States to recognize the need for restraint just as they restrain those countries that export to them.

Many companies in the world steel industry, including a number in Western Europe, are in desperate straights and are seeking almost any outlet to increase their steel production. This has placed the problem on a political level, not the ideal atmosphere in which to negotiate an economic agreement.

Unfortunately, during the talks on limitations and restraint, imports into the United States in June were almost 1.8 million net tons, of which 747,000 net tons came from the Common Market, an increase of 183,000 over the 564,000 net tons shipped in May. This has further charged the atmosphere in which the present negotiations are taking place.

Negotiations with the EEC on import restrictions are difficult because of the composition of that body. Although it is an economic community, the individual countries comprising it have their own objectives and goals. Unlike the Japanese steel industry, which has basically five large integrated companies, all Japanese, the EEC steel industry is made up of many more large integrated companies in seven different countries. Consequently, it is very difficult for any central body, even the Commission of the EEC, to exercise firm control. This was proven in the late 1970s and up through 1981 when the Davignon Plan and the subsequent imposition of Article 58 were not completely observed by all of the companies involved.

The latest offer to control exports from the EEC, made in August 1982, was termed unsatisfactory by the U.S. steel industry, since the tonnage figure was too high. It was not surprising that this did not represent much of a reduction from the high level of exports from the EEC to the United States in 1981, since all of the companies in all of the countries want to export as much as possible to alleviate their present economic distress. This is also true of many companies in other countries outside of the EEC that export steel to the United States.

Negotiations with the EEC will continue to be difficult because of the number of companies and the variety of interests involved. This was illustrated in the 1971–1972 negotiations on voluntary export quotas, when the Japanese were able to reach an agreement in a very short time, whereas negotiations with the EEC continued for many more months.

In the decade ahead, world steel-trade problems would be solved and the survival of current participants in the steel industry assured, if demand increases so that capacity can be used at the 1973–1974 rate. If, on the other hand, the 1982 conditions persist for another year, trade problems will intensify, and there will be a number of casualties among steel producers.

This is true of the industry in the United States for which 1983 will be an extremely crucial year. Losses in the second quarter of 1982 were staggering and there is not much hope for improvement in the third and fourth quarters. It has been estimated that the U.S. industry will lose upwards of $2 billion for the entire year. If 1983 is a repetition, a number of companies face disaster. Virtually every company has cut back on capital expenditures and adopted programs to reduce operating expenses in order to conserve cash. Decisions are pending on the reduction in size or the closure of a number of plants, and should the present depression continue, many of these will be made, reducing overall capacity by a substantial amount. In place of the 8 to 10 million tons which are currently designated for closure, the amount could well reach 15 to 16 million tons with several companies on the verge of bankruptcy, which might necessitate mergers as a means of survival.

If, on the other hand, the demand for steel and consequently the rate of industry operations improve, a number of companies that have sustained large losses will be able to recover. Thus, the next nine to twelve months will determine the fate of a significant segment of the U.S. steel industry. If the depression continues, only the strongest will survive.

Future operating rates in the world steel industry depend on the relationship between supply and demand. By 1985, world capacity will increase but slightly, however, in individual areas, such as the United States, the EEC, and, to a lesser extent, Japan, it will decline notably. Thus, with a shrinkage in supply, it will take but a small increase in demand to produce an increase in operating rates. Further, the supply, which will be withdrawn, consists of obsolete facilities, and those that remain will be more efficient. By 1985, with less steelmaking capacity available in a number of areas, a modest increase in demand above the 1979 level could produce a tight steel situation. Such a development, particularly with efficient equipment in operation, could well bring profitable results.

No matter what growth there is in demand, the sure survivors in the steel industry will be those private-sector companies that are efficient producers and particularly those that have diversified sources of income in addition to their steel operations. Government-owned and subsidized companies will also survive as long as they continue to be supported. Large, inefficient privately owned companies will be reduced in size or disappear as will some small private companies. Others with less than 1 million tons of capacity will find their survival in a localized market or in product specialization.

The survivors, although they may be somewhat reduced in size, will have most of their facilities modernized and efficient, if current capital-investment programs are carried out. Thus, the world steel industry in the mid- and late-1980s, generally speaking, should be a more efficient operation than it is in 1982.

A moratorium, until 1985, on the installation of added steel capacity throughout the world would be most welcome. An overbuilt steel industry with a considerable amount of obsolete equipment has been a major cause of the present predicament. A moratorium on the installation of new capacity, coupled with a modernization and replacement program, will contribute substantially to the solution.

Notes

1. Japan Steel Information Center, *Questions and Answers on the Japanese Steel Industry,* New York, July 1982, p. 84.

2. *American Metal Market,* June 25, 1982, p. 8.

3. Ibid., June 30, 1982, p. 9.

4. *American Metal Market,* June 23, 1982, p. 4.

5. American Iron and Steel Institute, *Steel Import News,* June 25, 1982.

Bibliography

Abegglen, James C. "U.S.-Japan Technological Exchange in Retrospect, 1946–1981. "In *Technological Exchange: the U.S. Japanese Experience, Proceedings of Japan-America Society of Washington Symposium, October 21, 1981.* Washington, D.C.: University Press of America, 1982.

Adams, Walter, and Dirlam, Joel B. "Big Steel, Invention, and Innovation." *Quarterly Journal of Economics.* 80, no. 2, May 1966.

———. "Steel Imports and Vertical Oligopoly Power." *American Economic Review.* 54, no. 5, September 1964.

American Enterprise Institute. *Fiscal Policy and Business Capital Formation.* Washington, D.C.: AEI, 1967.

American Iron and Steel Institute. *Annual Statistical Reports.* Washington, D.C.: AISI, various years.

———. *Directory of Iron and Steel Works of the United States and Canada.* Washington, D.C.: AISI, various years.

———. *Steel at the Crossroads: One Year Later.* Washington, D.C.: AISI, June 1981.

———. *Steel at the Crossroads: the American Steel Industry in the 1980s.* Washington, D.C.: AISI, 1980.

———. *Steel Industry Economics and Federal Income Tax Policy.* Washington, D.C.: AISI, June 1975.

———. *The Steel Import Problem.* New York: AISI, various years.

American Metal Market Co. *Metal Statistics: The Purchasing Guide of the Metal Industries.* New York: American Metal Market, various years.

American Petroleum Institute. *Basic Petroleum Data Book.* Washington, D.C.: API, 1979.

ARBED, S.A. *A Portrait of the Group.* Luxembourg: ARBED, 1979.

Baer, Werner. *The Development of the Brazilian Industry.* Nashville, Tennessee: Vanderbilt University Press, 1969.

Barnett, Donald. "Labor Productivity Trends in the U.S. Steel Industry." *Economic Papers.* Washington, D.C.: American Iron and Steel Institute, October 1980.

Battelle Memorial Institute. *A Survey and Analysis of the Supply and Availability of Obsolete Iron and Steel Scrap.* Columbus, Ohio: Battelle, 1957.

———. *Technical and Economic Analysis of the Impact of the Recent Developments in Steelmaking Practices on the Supplying Industries.* Columbus, Ohio: Battelle, 1964.

Beigie, Carl E., and Hero, Alfred O., eds. *Natural Resources in U.S.-Canadian Relations, Vol. II Patterns and Trends in Resource Supplies and Policies.* Boulder, Colorado: Westview Press, 1980.

Bieda, K. *The Structure and Operation of the Japanese Economy.* New York: John Wiley and Sons, Inc., 1970.

Blauvelt, W.W. *A Free Market: Ferrous Scrap Demand and Supply.* Atlanta, Georgia: Ferrous Scrap Consumers Coalition Symposium, February 4 and 5, 1980.

Boylan, Myles. *Economic Effects of Scale Increase in the Steel Industry.* New York: Praeger Publishers, 1976.

Bradford, Charles, A. *Japanese Steel Industry: A Comparison with its United States Counterpart.* New York: Merrill Lynch Securities Research Division, 1977.

British Steel Corporation, Statistical Services. *International Steel Statistics: World Tables.* Croydon, England: BSC, various years.

————. *Ten Year Development Strategy.* London, England: Her Majesty's Stationary Office, 1973.

Broken Hill Proprietary Co. Ltd., Public Affairs Department. *BHP Pocketbook.* Melbourne, Australia: BHP, 1977.

Brown, James W., and Reddy, Richard L. "Electric Arc Furnace Steelmaking with Sponge Iron." *ILAFA Direct Reduction Congress.* Macuto, Venezuela: Latin American Iron and Steel Institute, 1977.

Brunner, J.H.K.; Kelley, M.; Neath, D.S.; and Stewardson, B.R. *Raw Materials for the World Steel Industry.* Melbourne, Australia: Broken Hill Proprietary Co. Ltd., June 1977.

Burlingame, Richard D. "Trends in Scrap Quality for the 1980s." *Proceedings of the Electric Furnace Conference of the American Institute of Mining, Metallurgical and Petroleum Engineers, Toronto, Canada, December, 1978.* Warrendale, Pennsylvania: Iron and Steel Society of AIME, 1979.

Charles River Associates. *A Study of the Ferrous Scrap Market During the Shortage Period of 1973-1974.* Boston, Massachusetts: CRA, 1976.

Clark, David H. "Computer Process Control in the Steel Industry." *Report of Proceedings, Fifteenth Annual IISI Conference, Toronto, Canada, October 12-14, 1981.* Brussels, Belgium: International Iron and Steel Institute, 1982.

Cockerill, Anthony. *The Steel Industry: International Comparisons of Industrial Structure and Performance.* New York: Cambridge University Press, 1974.

Comptroller General of the United States. *Report to the Congress of the United States: New Strategy Required for Aiding Distressed Steel Industry.* Washington, D.C.: General Accounting Office, 1981.

Concast A.G. *World Survey of Continuous Casting Machines for Steel,* 6th ed. Zurich, Switzerland: Concast Documentation Center, 1980.

Cooper, Franklin D. "Potential Production Capacity of Metallurgical Coke." *Iron and Steel Engineer* 58 (September 1981):52–57.

Cordero, Raymond, and Serjeantson, Richard, eds. *Iron and Steel Works of the World,* 7th edition. London, England: Metal Bulletin Books, Ltd., 1978.

Council on Wage and Price Stability. *Report to the President on Prices and Costs in the United States Steel Industry.* Washington, D.C.: U.S. Government Printing Office, October 1977.

Crandall, Robert W. *The U.S. Steel Industry in Recurrent Crisis.* Washington, D.C.: The Brookings Institution, 1981.

Dancy, Terence E., and Laferrier, George H. "Large-Scale Integrated Steelmaking Using Direct Reduced Iron." *Proceedings of the 88th General Meeting of the American Iron and Steel Institute, New York, May, 1980.* Washington, D.C.: American Iron and Steel Institute, 1981.

Deily, Richard L., ed. "The World Steel Industry: Into and Out of the 1980s." *IISS Commentary.* Green Brook, New Jersey: Institute for Iron and Steel Studies, October 1981.

Dilley, David R., and McBride, David L. "Oxygen Steelmaking-Fact vs. Folklore." *Iron and Steel Engineer.* Pittsburgh, Pennsylvania: Association of Iron and Steel Engineers, October 1967.

European Business Reports Editorial Board. *The New EEC Measures for European Steel Production and Trading: From Crisis to Cartel?* European Business Reports No. 1. London, England: European Business Publications, 1981.

Eregli Iron and Steel Works. *Eregli 1980 Annual Report.* Ankara, Turkey: Eregli, 1981.

Estey, William Z. *Steel Profits Enquiry.* Ottawa, Canada: Information Canada, October 1974.

Federal Trade Commission, Bureau of Economics. *Staff Report on the United States Steel Industry and its International Rivals: Trends and Factors Determining International Competitiveness.* Washington, D.C.: U.S. Government Printing Office, November 1977.

Field, Lewis I. "Bethlehem Steel Corporation Experience with FIOR Direct Reduced Iron Briquettes." *11th Regular Meeting of the Committee on Technology of the International Iron and Steel Institute.* Rio de Janeiro, Brazil: April 1979.

Ford, Joseph W. *The Steel-Import Problem.* New York: Fordham University Press, 1961.

Gill, Dan R., and Kruppa, William S. *Iron Direct Reduction Process: Status and Description.* Washington, D.C.: U.S. Department of Commerce, 1976.

Gold, Bela. "Steel Technologies and Costs in the U.S. and Japan." *Iron and Steel Engineer Year Book, 1978.* Pittsburgh, Pennsylvania: Association of Iron and Steel Engineers, 1978.

Gold, Bela; Rosegger, Gerhard; and Boylan, Myles G. *Evaluating Technological Innovations.* Lexington, Massachusetts: Lexington Books, D.C. Heath and Company, 1980.

Hirschhorn, Joel S. "Continuing Success for United States Mini Mills." *Metal Bulletin's Second International Mini Mills Conference.* Vienna, Austria: March 7-9, 1982.

———. *Technology and Steel Competitiveness: Ferrous Scrap Trends and Issues.* Atlanta, Georgia: Ferrous Scrap Consumers Coalition Symposium, February 4 and 5, 1980.

Hogan, William T. *Economic History of the Iron and Steel Industry in the United States,* five volumes. Lexington, Massachusetts: Lexington Books, D.C. Heath and Company, 1971.

———. "Future Steel Plans in the Third World." *Iron and Steel Engineer* 54 (November 1977):25-37.

———. *Siderurgia Mundial: Perspectivas para la Decada del '80.* Buenos Aires, Argentina: Centro Internacional de Informacion Empresaria, 1981.

———. *The 1970s: Critical Years for Steel.* Lexington, Massachusetts: Lexington Books, D.C. Heath and Company, 1972.

———. "The Steel Import Problem: A Question of Quality and Price." *Thought.* 40, no. 159 (Winter 1965).

———. "World Steel in the 1980s." *Center Lines.* Cleveland, Ohio: Steel Service Center Institute, August 1981.

Hogan, Willam T., and Koelble, Frank T. *Analysis of the U.S. Metallurgical Coke Industry.* Washington, D.C.: U.S. Department of Commerce, 1979.

———. *Direct Reduction as an Ironmaking Alternative in the United States.* Washington, D.C.: U.S. Department of Commerce, 1981.

———. *Purchased Ferrous Scrap: United States Demand and Supply Outlook.* New York: Industrial Economics Research Institute, 1977.

Holschuh, Lenhard J. "Report of the Secretary General." *Report Proceedings, Fifteenth Annual IISI Conference, Toronto, Canada, October 12-14, 1981.* Brussels, Belgium: International Iron and Steel Institute, 1982.

Hoshii, Iwao. *Japan's Business Concentration.* Philadelphia, Pennsylvania: Orient/West Publishers, 1969.

Heuttner, David. *Plant Size, Technological Change, and Investment Requirements.* New York: Praeger Publishers, 1974.

Industrial Economics Research Institute, Fordham University. *Financial Study of the U.S. Steel Industry: Analysis of Capital Needs and Financial Condition.* New York: IERI, August 1975.

Innace, Joseph J. "Steel's Stampede into OCTG." *33 Magazine* 20 (March 1982):49-54.

Instituto Brasileiro de Siderurgia. *IBS Yearbook, Anuario Estatistico da In-*

dustria Siderurgica Brasilerira, 1981. Rio de Janeiro, Brazil: IBS, 1981.

International Iron and Steel Institute. *Report of Proceedings, Seventh Annual IISI Conference, Johannesburg, South Africa, October 7–11, 1973.* Brussels, Belgium: IISI, 1974.

International Iron and Steel Institute, Committee on Economic Studies. *Financing Steel Investment 1961–1971.* Brussels, Belgium: IISI, 1974.

————. *Projection 85: World Steel Demand.* Brussels, Belgium: IISI, March 1972.

————. *World Indirect Trade in Steel.* Brussels, Belgium: IISI, 1974.

International Iron and Steel Institute, Committee on Raw Materials. *Report on Iron Ore.* Brussels, Belgium: IISI, April 1977.

International Iron and Steel Institute, Committee on Statistics. *Steel Statistical Yearbook 1981.* Brussels, Belgium: IISI, 1981.

International Iron and Steel Institute, Committee on Technology. *A Study of the Continuous Casting of Steel.* Brussels, Belgium: IISI, 1977.

International Monetary Fund. *World Economic Outlook,* Occasional Paper No. 9. Washington, D.C.: IMF, 1982.

Jaffe, John, W.P. "International Cooperation in the Planning, Financing, Construction, and Operation of Iron and Steel Plants." *United Nations Third Interregional Symposium on the Iron and Steel Industry.* Brasilia, 1973.

Japanese Automobile Manufacturers Association. *Motor Vehicle Statistics of Japan 1981.* Tokyo, Japan: JAMA, 1981.

Japan Iron and Steel Federation. *Monthly Report of Iron & Steel Statistics.* Toyko, Japan: JISF, various monthly issues.

————. *Statistical Yearbook.* Toyko, Japan: JISF, various years.

————. *The Steel Industry of Japan 1973.* Toyko, Japan: JISF, 1973.

Jensen, Harold B. *New Alternatives for Charge Metallics.* Atlanta, Georgia: Ferrous Scrap Consumers Coalition Symposium, February 4 and 5, 1980.

Kawahito, Kiyoshi. *The Japanese Steel Industry: With an Analysis of the U.S. Steel Import Problem.* New York: Praeger Publishers, 1972.

Kawata, Sukeyuki, ed. *Japan's Iron and Steel Industry.* Toyko, Japan: Kawata Publicity, Inc., various years.

Korf, Willy. "Ways of Overcoming the Crisis in the European Steel Industry." *Paper presented to the 67th German Entrepreneurs' Seminar.* Baden-Baden, West Germany: October 1981.

Kruppa, William S., and Palmer, K.W. *Production of Iron by Direct Reduction.* U.S. Department of Commerce, Industry and Trade Administration. Washington, D.C.: U.S. Government Printing Office, 1979.

Labee, C.J., and Samways, N.L. "Developments in the Iron and Steel Industry, U.S. and Canada—1981." *Iron and Steel Engineer* 59 (February 1982):D1–D34.

Liedholm, Carl. *The Indian Iron and Steel Industry: An Analysis of Comparative Advantage.* East Lansing, Michigan: MSU International Business and Economics Studies, 1972.

Malenbaum, W. *World Demand for Raw Materials in 1985 and 2000.* Philadelphia, Pennsylvania: University of Pennsylvania, October 1977.

Manners, Gerald. *The Changing World Market for Iron Ore 1950–1980.* Baltimore, Maryland: Johns Hopkins Press, 1971.

Marcus, Peter F. *Internationalization of Steel.* New York: Mitchell Hutchins, Inc., February 1974.

————. *World Steel Supply Dynamics: 1975–1981.* New York: Mitchell Hutchins, Inc., March 1976.

Mastey, Anthony D. "Productivity: People are the Margin of Difference." *33 Magazine* 19 (March 1981):45–52.

Mazel, Joseph L. "Importing Steel vs. Importing Technology: How American is American Steel?" *33 Magazine* 20 (May 1982):70–75.

Metal Bulletin Ltd. *The Iron and Steel Industry in China.* Surrey, England: Metal Bulletin, 1978.

McAdams, Alan K. "Big Steel, Invention, and Innovation, Reconsidered." *Quarterly Journal of Economics.* 81, no. 3 (August 1967).

McBride, D.L., and Dancy, T.E., eds. *Continuous Casting.* New York: John Wiley & Sons, 1962.

McManus, George, "Auto Market Demand Has Sheet Production Stalled in First Gear." *Iron Age* 225 (May 3, 1982):MP3–MP11.

Merrill, John P.; Pifer, Howard W.; Marshall, Paul W.; and Breitenberg, Matthew D. *The Economic Implications of Foreign Steel Pricing Practices in the U.S. Market.* Newton, Massachusetts: Putnam, Hayes and Bartlett, Inc., August 1978.

Monnet, A.A. "Perspectives on Protectionism." *Iron and Steel Engineer Yearbook, 1978.* Pittsburgh, Pennsylvania: Association of Iron and Steel Engineers, 1978.

Motor Vehicle Manufacturers Association. *World Motor Vehicle Data.* Detroit, Michigan: MVMA, various years.

Mueller, Hans. "Present and Future Problems in the U.S. Steel Market." *Annual Conference of the Association of Steel Distributors.* New York: December 1981.

————. *The Competitiveness of the U.S. Steel Industry after the New Trigger Price Mechanism.* Murfreesboro, Tennessee: Middle Tennessee State University, Monograph Series No. 25, Business and Economic Research Center, December 1980.

Nakajima, Yuji, ed. *Problems of Transportation in Japan.* Tokyo, Japan: Institute of Transportation Economics, 1978.

National Commission on Supplies and Shortages. *Government and the Nation's Resources.* Washington, D.C.: U.S. Government Printing Office, December 1976.

Organisation for Economic Cooperation and Development. *Emission Control Costs in the Iron and Steel Industry.* Paris, France: OECD, 1977.
———. *Investment in the Iron and Steel Industry, Problems and Prospects.* Paris, France: OECD, October 1974.
———. *The Iron and Steel Industry.* Paris, France: OECD, various years.
Packard, Ruby, ed. *Metal Bulletin Handbook.* London, England: Metal Bulletin Books, Ltd., various years.
Pifer, Howard W.; Marshall, Paul. W.; and Merrill, John P. *Economics of International Steel Trade: Policy Implications for the United States.* Newton, Massachusetts: Putnam, Hayes and Bartlett, Inc., 1977.
Robert Nathan Associates. *Iron and Steel Scrap.* Washington, D.C.: Metal Scrap Research and Education Foundation, 1977.
Rosegger, Gerhard. "Steel Imports and Vertical Oligopoly Power: Comment." *American Economic Review.* 57 no. 4, September 1967.
Ruhrkohle Aktiengesellschaft. *Figures, Data, Facts.* Essen, West Germany; Ruhrkohle, 1980.
Russell, C., and Vaughn, W. *Steel Production: Processes, Products, and Residuals.* Baltimore, Maryland: Johns Hopkins University Press, 1976.
Schneider, Phillip E., *Documented Japanese Steel Production Costs, Prices, and Business Procedures.* Pittsburgh, Pennsylvania: International Ventures Management, Inc., 1977.
Schoenbrod, D.S., and Hone, G.A. "Steel Imports and Vertical Oligopoly Power: Comment." *American Economic Review.* 56, no. 1, March 1966.
Sheridan, Eugene T. *Supply and Demand for United States Coking Coals and Metallurgical Coke.* Washington, D.C.: U.S. Department of the Interior, Bureau of Mines, 1976.
Slesinger, R.E. "Steel Imports and Vertical Oligopoly Power: Comment." *American Economic Review.* 56, no. 1, March 1966.
South African Iron and Steel Industrial Corporation Ltd., *1980 Annual Report.* Pretoria, South Africa: ISCOR, 1981.
Stanford Research Institute. *Investment Outlook and Related Federal Policies for the Steel, Plastic, and Paper Industries, 1976–1985.* Menlo Park, California: SRI, January 1977.
Stephenson, Robert L., ed. *Direct Reduced Iron: Technology and Economics of Production and Use.* Warrendale, Pennsylvania: Iron and Steel Society of AIME, 1980.
Szekely, Julian, ed. *Blast Furnace Technology: Science and Practice.* New York: Marcel Dekker, Inc., 1972.
33 Magazine. World Steel Industry Data Handbook, Vol. 1, The United States. New York: McGraw-Hill Inc., 1978.
United Nations, Economic Commission for Europe. *Aspects of Competition between Steel and Other Materials.* New York: United Nations, 1966.

————. *Long Term Trends and Problems of the European Steel Industry.* Geneva, Switzerland: United Nations, 1959.

————. *Quarterly Bulletin of Steel Statistics for Europe.* New York: United Nations, various years.

————. *Statistics of World Trade in Steel.* New York: United Nations, various years.

————. *The Steel Market.* New York: United Nations, various years.

————. *The World Market for Iron Ore.* New York: United Nations, 1968.

United Nations, Industrial Development Organization. *Draft World-Wide Study of the Iron and Steel Industry: 1975–2000.* Vienna, Austria: International Centre for Industrial Studies, December 1976.

————. *Proceedings of the Second General Conference.* Lima, Peru: UNIDO, 1975.

United States Congress, House of Representatives. *Crises in the Steel Industry: An Introduction and the Steel Industry in Transition.* 97th Congress, 2nd Session, Committee Print 97–EE. Washington, D.C.: U.S. Government Printing Office, 1982.

————. *Message from the President of the United States Relative to our Federal Tax System.* 87th Congress, 1st Session, April 20, 1961. House Document No. 140. Washington, D.C.: U.S. Government Printing Office, 1961.

United States Congress, Office of Technology Assessment. *Technology and Steel Industry Competitiveness.* Washington, D.C.: U.S. Government Printing Office, 1980.

United States Department of Labor, Bureau of Labor Statistics. *An International Comparison of Unit Labor Cost in the Iron and Steel Industry, 1964: United States, France, Germany, United Kingdom,* Bulletin 1580. Washington, D.C.: U.S. Government Printing Office, 1968.

United States Department of the Interior, Bureau of Mines. "Iron and Steel Scrap, Monthly." *Mineral Industry Surveys.* Washington, D.C.: U.S. Government Printing Office, various monthly issues.

————. *Minerals Yearbook.* Washington, D.C.: Bureau of Mines, various years.

United States Department of the Treasury, Office of Tax Analysis. *Federal Tax Policy and Recycling of Solid Waste Materials.* Washington, D.C.: U.S. Department of the Treasury, February 1979.

United States Senate, Committee on Finance. *Steel Imports.* 90th Congress, 1st Session, December 19, 1967. Washington, D.C.: U.S. Government Printing Office, 1967.

United States Steel Corporation, Monroeville Research Laboratory. *Compilation of Reports on Direct Reduction Technology and Economics.* Idaho Falls, Idaho: U.S. Department of Energy, 1981.

United States Steel Industry Task Force. "A Comprehensive Program for the Steel Industry," *Report to the President of the United States.* Washington, D.C.: December 13, 1977.

Vaugn, W.; Russell, C.; and Cochran, H. *Government Policies and the Adoption of Innovations in the Integrated Iron and Steel Industry.* Washington, D.C.: Resources for the Future, 1976.

Warren, Kenneth. *World Steel: An Economic Geography.* New York: Crane Russak, 1975.

Watkins, Ralph J.; Fuller, Kemp G.; and Fisher, Jacob. *The Market for Steel in Mexico.* New York: Praeger Publishers, 1964.

Woods, Gordon, et al. *1980–90 Forecast of U.S. Ferrous Scrap Prices.* Hamilton, Canada: Woods, Gordon, July 1979.

Zapata, Francisco. *Las Truchas, Acero y Sociedad en Mexico.* Camino al Ajusco, Mexico: El Colegio de Mexico, 1978.

Zinder-Neris, Inc. *Coal Statistics International.* New York: McGraw-Hill, Inc., various monthly issues.

Index

Acindar-Industria Argentina de Aceros SA, 170, 171, 172
Africa, 8, 33, 34, 35, 51, 69, 75, 98, 100
Alan Wood Steel Company, 91, 93, 201, 202
Algeria, steel industry in, 175, 189
Algoma Steel Corporation, 137, 138, 139
Alphasteel, 25
Altos Hornos de Mexico SA, 166, 167, 168
Altos Hornos de Vizcaya SA, 56
Altos Hornos del Mediterraneo SA, 56
Altos Hornos Zapla, 171
American Iron and Steel Institute, 96, 197, 201, 234
Anshan Works, 157
Antidumping Act, U.S.:
  enforcement of, 203, 236, 237
  suits filed under, 116, 118, 130, 202, 203, 204, 207, 236
  Treasury Department, and, 203
  trigger-price mechanism, and, 203, 207, 237
AOD process, 45
Arbed SA, 25, 28, 29, 31, 35, 38, 39, 46, 49, 102, 161
Argentina, steel industry in, 5, 8, 11, 85, 170, 171, 172, 194, 195, 216
Argon oxygen decarburization, 12
Armco Inc., 91, 95, 99, 110, 111, 117, 140
Article 58, EEC treaty of, 21, 22ff
Athlone Industries, 113
Atlantic Steel Company, 122
Atlas Steel Company, 138, 139
Australia:
  Brazil, competition with, iron-ore market, for, 148
  coking-coal exports of, 38, 40, 70, 72, 146
  coking coal in, 9, 146, 224
  iron-ore discoveries in, by U.S. steel industry, 98
  iron-ore exports of, 33, 34, 35, 37, 51, 69, 75, 76, 146, 147, 148
  iron-ore in, 8, 33, 98, 146, 148, 224
  iron-ore transportation, EEC, to, 37
  steel industry in, 144-148
  trade, international steel, in, 85, 144-145, 216
Australia, steel industry in:
  Armco, proposed joint venture of, 147
  basic-oxygen steelmaking in, 146, 147
  blast furnaces of, 145, 147
  Broken Hill Proprietary, 144, 148
  capacity of, 144, 145, 222
  capital investment, future, in, 146
  closures, plants, imports, and, 145
  continuous casting in, 146, 147

costs in, 145
electric-furnace steelmaking in, 147
finishing-mill companies of, 144
hot-strip mill, new, in, 146
Jervis Bay, proposed joint venture at, 147
John Lysaght, 146
joint ventures proposed in, 147-148
Kaiser Steel, proposed joint venture in, 147
Kwinana blast-furnace plant, 144, 145
minimill, at Melbourne, 144
modernization in, 146
Mount Newman Project, 148
Newcastle Works, 144, 145
open-hearths of, 145-147
ownership, of private, 5, 145, 216
outlook for, 148, 222
Port Kembla plant, 144, 145, 146
production of, steel, 144
productivity in, 145
raw materials for, 146-148
Western Australia, proposed joint venture in, 147-148
Whyalla plant, 144, 145, 147
Austria:
  basic-oxygen furnace in, introduction of, 41, 54, 55, 107
  population of, steel and, 54
  steel industry in, 41, 54-55
  trade, international, steel, in, 48, 53, 55
Austria, steel industry in:
  Alpine, origins of, 54
  basic oxygen steelmaking process: first use of, by, 54, 55, 107; furnaces replaced by, 54
  blast-furnace replacement by, 54
  capacity, expansion plans, lack of, 55
  coking coal for, imports of, 55
  continuous casting in, 53
  Donawitz, integrated plant, Alpine, of, at, 54
  EEC, steel exports to, 55
  efficiency of, future, in, 51
  electric-furnace companies in, 54
  iron-ore, imports of, for, 55
  Linz, integrated plant at, 54, 107
  mergers in, 54
  open-hearth replacements by, 54
  outlook for, 55
  ownership of, public, 55
  production, steel, of, 52, 54, 55
  raw materials for, 53, 55
  Schoeller-Bleckmann, absorbed, Voest-Alpine, by, 54
  size of, 51, 54

250

steelmaking processes used by, 52, 53, 54
technology, contributions to, by, 55
Voest, formation of, 54
Voest-Alpine, 54, 55
Automobile industry:
downsizing in, 127
Japan, in, 3, 4, 16–17, 83, 126, 210
petroleum, demand, and, 3
production decline in, 1975, 17
registrations, automobile, 3
shipments, steel, to, United States, in, 124–125, 126
steel demand, and, 3, 16, 124–125, 126, 210, 212
United States, in, 3, 17–18, 124–127, 142, 212
Western Europe, in, 3
worldwide, 3

Baldrige, Malcolm, U.S. Secretary of Commerce, 205
Bangladesh, steel industry projects in, 175
Bankruptcies, steel company:
Alan Wood Steel, of, 93
Bresciani, prospective, of, 30, 50
government assistance, and, 1, 49
Italy, prospective, in, 30, 50
reorganization, steel industry of, and, 219
trade, international, steel, and, 7, 201
United States, in, 7, 18, 91, 93, 96, 113, 240
Western Europe, in, 1, 18, 19, 20, 50
Wisconsin Steel, of, 93
Baoshan Steel Works, 156, 158, 159
Basic-oxygen steelmaking process. See Technology, steel
Bayou Steel Corporation, 122
Belgium:
coal imports of, 38
coal production in, 38
government of, 29, 30, 31, 46
iron-ore, imports of, 34
steel industry in, 28–29, 30–31, 46
trade, international, steel, in, 193
Belgium, steel industry in:
Arbed, relationship, Sidmar, to, 31, 46
basic-oxygen steelmaking in, 42, 44
Boel, 30, 31, 46
capacity of, 31, 221
Clabecq, 30, 31, 46
Cockerill, 30, 31, 46, 205, 212
composition of, 30
continuous casting in, 44
countervailing duties on, U.S., by, 205, 212
difficulties in, 30, 46
direct-reduced iron, demand for, in, 45
electric-furnace steelmaking in, 41, 42
government role in, 5, 29, 31, 46, 49
integrated companies in, 30
Japanese study of, 30
merger, proposed in, 30, 46
modernization requirements of, 31
ownership, government, in, 30, 46, 49
plant, most modern, 28, 31
political problems, and, 31
production, steel, in, 192, 196
raw materials for, 33, 34, 38, 41, 45
reorganization of, 30–31, 46
Sambre, 30, 31, 46, 205, 212
scrap demand in, 41, 45
Sidmar, 30, 31, 46
specialty producers in, 30
subsidization of, 5, 46, 49, 205
Bethlehem Steel Corporation, 91, 93, 94, 95, 96, 97, 99, 117, 119, 120, 121, 124, 140, 202, 227
Blast furnaces. See Technology, steel
Boel, Usines Gustave Boel SA, 30, 31, 46
Boom, steel:
automobile production, decline of, and, 16–17
bankruptcies, after, 18
Canada, steel industry in, and, 137
capacity limitations, and, 15
capital spending, decline of, and, 17
coking coal prices, and, 9
demand, steel, during, 15–16
end of, 4, 15–18
energy crisis, and, 16
growth outlook during, 15, 92–93
height of, 1
India, steel industry in, and, 163
inventories, steel, excessive, and, 17
Japanese steel industry, and, 15–18, 63
losses, after, 18
oil prices, and, 16
optimism, steelmakers, of, and, 15
prices, steel, during, 16, 199
production, raw steel, and, 15, 17, 33
profits, steel, and, 15, 18
raw-material limitations, and, 15
shortages, steel, during, 16
Third World, and, 18
trade, international, steel, and, 16, 18, 199, 200, 207, 210, 230
United States Steel industry, and, 15–18, 92, 128, 129, 199
Brazil:
Australia, competition with, iron-ore market, for, 148, 161
Carajas, iron-ore deposit at, 223
coking coal, imports of, 8, 102, 163
countervailing duties on, U.S., by, 205
EEC, Carajas financing, and, 224
Export-Import Bank of Japan, Carajas, financing, and, 224
export market, steel, Japan, for, 84

iron-ore discoveries in, U.S. steel
    industry, by, 98
iron-ore exports of, 33, 34, 35, 37, 51,
    75, 100
iron ore in, 8, 33, 98, 161, 163, 223, 224
Japan, Carajas loan of, 224
joint ventures, iron ore, in, 34
OECD International Steel Committee,
    nonmembership in, 215
steel industry in, 155, 161–163, 176, 177,
    178, 179, 229
trade, international, steel, in, 85,
    162–163, 177, 194, 195, 205, 213, 215,
    216, 231, 232
United States, steel market in, and, 231
World Bank, Carajas financing, and,
    223–224
Brazil, steel industry in:
Acominas plant, 162
Arbed, involvement in, 161
basic-oxygen steelmaking in, 162
Belgo-Mineira, 161
Bessemer steelmaking in, 162
Brazilian Iron and Steel Institute, 161, 231
capacity expansion by, 161, 162, 229
charcoal, use of, blast furnaces in, 162
continuous casting in, 162
Cosipa, 161
CSN, 161, 162
CVRD, Acominas plant, participation in,
    162
electric-furnace steelmaking in, 162
export market, steel, and, 213
greenfield-site plants, construction of,
    161–162
integrated producers in, 161
Italsider, Tuberao participation of, 162
joint ventures'in, 162
Kawasaki, Tuberao participation of, 162
Mannesmann, 161
Minas Geros, state of, Acominas plant,
    and, 162
open-hearth steelmaking in, 162
outlook for, 161, 163
ownership, in, 5, 161, 162, 229
production, steel, in, 161, 162, 220
Siderbras, 162
Tuberao plant, 162
Usiminas, 161
Volta Redonda plant, 162
Bresciani, 30, 50
British Steel Corporation, 5, 21, 24, 25,
    39, 45, 140, 143, 205, 212, 221, 228,
    234
Brock, William E., U.S. Trade
    Representative, 205
Broken Hill Proprietary Company, Limited,
    144, 145, 146, 147, 148
Bulgaria, steel industry in, 183, 185, 189
Burlington Steel, 139

Canada:
automobile treaty, United States, with,
    142
coking coal in, 9, 102
Fire Lake Project, 99, 140, 143
Hollinger Mines Ltd., 140
iron ore in, 33, 34, 35, 51, 69, 98–99, 100
Iron Ore Company of Canada, 99, 140
Japan, raw materials shipped to, 69, 75
Labrador Mining & Exploration Co.
    Ltd., 140
Port Cartier, 140, 143
Quebec-Cartier Mines, 99, 140
raw materials, steelmaking, in, 139–140
steel industry in, 137–144
trade-international, steel, in, 85, 141–143,
    198
voluntary import restraints,
    nonparticipation in, 198
Wabush Mines, 99, 140
Canada, steel industry in:
Algoma Steel, 137, 138, 139
Atlas Steel, 138, 139
basic-oxygen steelmaking in, 138, 140,
    141
Big Three companies in, 137
boom, steel, and, 137
Burlington Steel, 138, 139
capacity of, 138, 141, 222
capital investment in, 138, 142, 143, 144
composition of, 137
continuous casting in, 141
direct reduction in, 11, 140, 141
Dofasco, 137, 138, 139, 140
DRI use in, 138
earnings in, 2, 137
efficiency of, 137
electric-furnace steelmaking in, 137,
    140, 141
employment in, 137
government policies, and, 143–144
growth of, 137
Interprovincial Steel, 138, 139
Ivaco, 139
Lasco, 139, 141
location of, 138
Manitoba Rolling Mills, 139
Midrex direct-reduction process in, 141
minimills in, 137, 138
open-hearth steelmaking in, 140, 141, 143
operating rates in, 137
outlook for, 138, 142, 222
ownership of, 5, 137, 144
production of, steel, 2, 137, 220
public companies of, 138, 143, 144
raw materials for, 139–140, 143
scrap use in, 138
Sidbec-Dosco, 137, 138, 139, 141, 143,
    144
Slater Steel, 139,

SLRN direct-reduction process in,
141
Stelco, 137, 138, 139, 140, 141
Sydney, 137, 138, 139, 141, 143, 144
technology in, 140–141
trigger prices, United States, and, 142
Western Canada Steel, 139
World War II, development after, 137,
139
Capacity, steel, expansion of:
Argentina, in, 170–171
Austria, lack of, in, 55
Australia, in, 144, 145
Belgium, in, prospects for, 29
boom, steel, and, 15–16
Brazil, in, 161, 162
Broken Hill Proprietary, by, 145
Canada, in, 138, 222
China, in, 155–159, 160, 222
Czechoslovakia, in, 188
East Germany, in, 189
electric furnaces, 9
Far East, in, 222
financing of, 15
French steel industry, in, 28
Hoogovens, lack of, in, 31
India, in, 164
Indonesia, in, 209, 222
Iscor, lack of, in, 149
Japan, in, xv, 2, 6, 16–17, 63, 64–65, 67,
68, 88, 222
Korean War, and, 106
Mexico, in, 167
Netherlands, lack of, in, 31
Non-EEC Western European countries,
in, 53
plans for: cancellation of, Davignon plan,
by, 20–21; moratorium on, proposed,
238; 1974, in, 16, 17, 21, 92–93, 220,
221, 222–223
Pohang Steel Company, in, 165
Poland, in, 188
South Africa, lack of, in, 149, 222
South Korea, in, xv, 165, 222
Soviet Union, 187
steel demand, and, 4, 220
Sweden, cancellation of, 58
Taiwan, in, 169, 222
Third World, in, xv, 2, 153, 175, 176,
213, 222, 223
trade, international, steel, growth of,
and, 2, 6, 191, 194, 208, 211
Turkey, in, 60
United States in, 1, 16–17, 92–93, 94,
97, 105–106, 117, 221
Venezuela, in, 172, 173
Voest-Alpine, lack of, in, 55
West German steel industry, lack of, in,
25
world, in, 1, 2, 219–220

Capacity, steel, reduction of:
Alan Wood Steel, and, 93, 202
Arbed, by, 29
Belgium, in, 31
Bethlehem Steel, by, 93, 119, 202
British Steel Corporation, by, 24–25, 221
Cockerill, proposed at, 31
Crucible Steel, by, 93
Cyclops, by, 93
Davignon plan, under, 20–21, 23ff, 92
EEC, future, in, 51, 221, 237
France, in, 27
Italy, lack of, in, 30
Japan, in, 63, 237
Kaiser Steel, by, 94, 119, 120
National Steel, by, 93, 94, 229
Peine-Salzgitter, by, 27
Sacilor, by, 27
Sambre, proposed at, 31
Thyssen, by, 25–26, 230
United States, in, 1, 92, 93, 94, 117, 119,
132, 202, 211–212, 221, 237
United States Steel, by, 93, 229
Usinor, by, 27
Western Europe, in, xv, 20–21, 51, 221,
237
Wisconsin Steel, and, 93
world steel industry, in, 219
Youngstown Sheet and Tube, by, 93, 202
Capital investment, steel:
Arbed, Saar, in, 29
Armco, postponement of, by, 123, 227
Australia, in, 146
Bethlehem Steel, by, 124, 227
British Steel, of, Teesside, at, 24
Canada, in, 138, 142, 143, 144, 227
China, projected, in, 159
government support for, EEC, in, 51, 222
Japan, in, 2, 63, 64, 67, 88, 211, 222
ownership, steel-company, and, 227
planned, 1974, 15–17
pollution control, for, 45
South Africa, in, 149
Sweden, in, 58
subsidies, and, 27, 29, 50
Sydney Steel, by, 227
technology, new, for, 226, 227
Third World, in, 222
United States, in, 94, 95, 97, 98, 101,
123–124, 131, 227, 231
United States Steel, by, 227
Carter, President Jimmy, 202, 203, 204
CF&I Steel Corporation, 92, 94, 113
Chile, steel industry in, 75, 174, 176
China, 71, 83, 154–161, 195, 206, 213
China Steel Company, 168, 169
China, steel industry, in:
Anshan Works, 157, 158
backyard furnaces of, 154, 159
blast furnaces in, 156, 157

Baoshan Steel Works, 156, 158, 159
capacity of, 156, 157, 158, 159
capital investments, in, projected, 159
Chitung Steel Works, 156–157, 158
Dresdner Bank, steel projects,
    participation in, 157
expansion plans, postponement of,
    158, 160
Fifth National People's Conference,
    and, 158
financing plans, expansion, for, 158
Japan, trade agreement with, 1978,
    in, 155
Mannesmann, steel projects for, 156
modernization of, 157, 159
Nippon Steel, steel projects for, 155–156
outlook for, 159, 160–161
production, steel, in, 154–155, 158
public ownership in, 5, 230
raw materials for, 157
Soviet assistance to, withdrawal of, 154
Thyssen, steel projects for, 157
United States Steel, project negotiations
    with, 157
World War II, changes since, 154
Wuhan Iron and Steel, 156
Chitung Steel Works, 156, 157, 158
Cia Siderurgiea Belgo-Mineira, 161
Cia Siderurgiea Mannesmann, 161
Clabecq, SA Forges de Clabecq, 30, 31, 46
Closures, plant, steel:
    Alan Wood Steel, 91, 201–202
    Arbed, by, 29
    Australia, in, 145
    Bethlehem Steel, by, 97, 117, 202
    British Steel Corporation, by, 24
    Danish steel industry, in, 32
    Duport Steel Limited, 25
    EEC, in, 51
    French steel industry, in, 27
    losses, financial, and, 1, 201–202
    mills at Dortmund, Hoesch, by, 25
    Patent Shaft Steel Works, 25
    Round Oaks Steel Works, 25
    Southeast, United States, in, 120
    United States, in, 91, 93, 97, 117, 119
        120, 130, 201, 202
    West Coast, United States, in, 120
    Wisconsin Steel, 91
Coal Arbed, 39, 102
Coal, coking. See Raw materials,
    steelmaking for
Coal Products Limited, 39
Cockerill, SA Cockerill-Ougree-Providence
    et Esperance-Longdoz, 30, 31, 46, 205,
    212
Coke ovens. See Technology, steel
Columbia, steel industry in, 176
Colt Industries, 113
Columbia Steel Company, 113

Competition, steel, in, xv, 12, 23, 49, 67,
    84–85, 191–216, 222, 227, 235
Computers, steel in, 10, 12, 45, 80, 112
Consumers Union of the United States,
    198–199
Continental Steel Corportation, 113
Continuous annealing, 12, 63, 80, 113
Continuous casting, steel. See Technology,
    steel
Cosipa-Cia Siderurgiea Paulista, 161
Crane Company, 113
Creusot-Loire, 28, 46
Crucible Steel Company, 91, 93, 113
Crude-oil production, 3, 4
CSN-Cia Siderurgiea Nacional, 161, 162
CVRD-Cia Vale do Rio Doce, 162
Cyclops Corporation, 91, 93
Czechoslovakia, steel industry in, 48, 55,
    185, 186, 187, 188

Dalmine Siderca Saic, 171
Dalmine SpA, 30
Davignon, Etienne, Viscount, 20, 21, 50
Davignon plan, 20, 21, 22, 23–32, 92
Demand, steel:
    Australia, in, Broken Hill Proprietary,
        and, 144–145
    automobile production, and, 3, 124–127
    Canada, in, steel expansion, and, 138
    capacity expansion, and, 219–220
    capital-goods demand, and, 4
    EEC capacity, and, 23, 51
    energy programs, United States, and, 132
    Japanese capacity, and, 63, 67, 208
    growth projections for, 4, 51
    mid-1950s, growth after, 2
    1960s and 1970s, during, 2, 4
    oil tankers, and, 3
    operating rates, and, 238
    pipelines, and, 3
    price concessions, and, 6
    satisfying of, xv
    shipbuilding, and, 3
    South Korea, in, 177
    supply, relation to, 6
    trade, international, steel, and
        fluctuations in, 6, 238
    uncertain nature of, xv, 1, 220
    United States, in, 92, 95, 97, 105–106,
        114, 127–128, 132
    Western Europe, in, supply, and, 28, 51
    world, in, 51, 63, 220, 238
Denmark, steel industry in:
    capacity of, 32, 40
    continuous casting in, 32, 44, 77
    Det Danske Stålvalsevaerk, 32
    electric furnaces of, 32, 40, 42
    exports, steel, by, 193
    open-hearth furnaces, closing of, 32
    ownership of, 46, 47

plates, produced by, 32
production, steel, peak, of, 32
raw materials for, 40
scrap, dependence on, 32, 40
Desulfurization, iron, 12
Det Danske Stålvalsevaerk AS, 32
Detroit Steel Corporation, 91
Direct reduction. *See* Technology, steel
Diversification, steel industry:
  Japan, in, 88
  management philosophy, and, 220–221
  National Steel, by, 229–230
  profitability, improved, and, 221
  survival strategy, and, 229
  United States, in, 2, 133, 221
  United States Steel, by, 229
Dofasco, Inc, 137, 138, 139, 140
Dongkuk Steel Mill Company Limited, 165
Dresdner Bank, 157
Dunswart Iron & Steel Works Ltd,
  148, 149
Duport Steel Limited, 25

Earnings, steel:
  boom, steel, and, 15, 18
  British Steel Corporation, restoration
    of, by, 25
  Canada, in, 2, 137
  diversification, and, 221
  EEC, in, 23, 48, 49, 50, 51
  Japan, in, 1, 15, 18
  United States, in, 1, 7, 15, 18, 91, 95,
    96–97, 129, 132, 201, 202, 203, 221
  Western Europe, in, 1, 15
East Germany, steel industry in, 34, 183
  184, 185, 189
Eastern Europe, steel industry in xv, 5, 34,
  48, 183, 184–185, 186, 187, 189, 227
EBV, subsidiary, Arbed, of, 39
EEC Commission, 220, 233
Egypt, steel industry in, 175, 176
Electric furnaces. *See* Technology, steel
Employment, steel:
  Belgium, in, 51
  Belgium-Luxembourg, in, 51
  British Steel, reduction in, 24, 25
  Canada, in, 137
  EEC, decline of, in, 51
  France, in, 51
  Italy, in, 51
  loss of, United States, in, 7, 119, 127,
    129, 202, 234
  maintenance of, 2, 5, 7, 191, 204, 206,
    228
  ownership, public, and, 5, 228
  prices, steel, and, 7, 228, 234
  subsidies, government, and, 5, 7, 228
  trade, international, and, 2, 5, 7, 191
    204, 206
  trade restrictions, and, 7

United Kingdom, in, 51
Venezuela, in, 172, 174
West Germany, in, 51
Energy crisis, 16–17, 30, 50, 75
Energy conservation, 1, 2, 63, 75, 80
Ensidesa-Empresa Nacional Siderurgica SA,
  56
Environmental Protection Agency, 102, 203
Establecimientos Metalurgicos Santa Rosa
  SA, 171
Estel. *See* Hoesch Werke AG
Eurofer I, 21
Eurofer II, 22, 23
European Economic Community:
  Article 58, treaty of, and, 21
  blast furnaces, shutdowns of, in, 41, 51
  capacity, steel, reduction in 24, 30, 51,
    92, 221, 237
  Carajas, iron-ore project, financing of,
    224
  coal imports, into, outlook, for, 40, 101
  coking coal imported by, 37–40
  Commission of, 220, 233
  continuous casting in, 43, 44, 51, 221,
    225
  costs, transportation, iron ore, of, 35, 37
  crisis, steel, in, 19ff, 49–51
  Davignon, Etienne, Viscount,
    Commissioner of, 20, 21
  Davignon plan, in, 20ff, 92
  direct-reduced iron, demand for, in,
    44–45
  direct reduction in, 44, 45
  employment, steel, decline in, 51
  Eurofer I, and, 21
  Eurofer II, and, 22ff
  European Independent Steel Association,
    and, 22
  freight rates, iron ore, to, 35–37
  French steel industry, ranking in, 25
  import restrictions, in, 7, 20, 48, 49, 50
  iron ore imported by, 33, 34, 35, 36,
    37, 51
  iron-ore production in, 32–33, 34
  Italian steel industry, in, 30, 50
  Japan, orderly marketing agreement,
    steel, with, 49, 200, 207
  joint ventures, iron ore, in, 34
  losses, financial, steel industry, in, 1, 18,
    19, 20, 21, 49, 50, 96, 97, 228
  mergers, steel, in, 23, 25, 26, 27, 51
  modernization, steel, in, 23–32, 49–51
  outlook, steel industry, for, 49–51
  ownership, steel industry, of, changes
    in, 45–47, 49–51, 216
  peripheral steel markets of, 49, 200,
    207, 211
  price concessions, steel, in, 7, 19, 48,
    49
  price cuts, Article 58, under, 22

price cutting, trade, steel, and, in, 48,
    49, 50
price restrictions, trade, on, 48, 207
price structure, steel, in, 1, 7, 19, 20
    22, 23, 49
prices, steel, increase in, 23, 50
problem, basic, steel industry of, in,
    49–51
production, steel, in, 1, 17, 19, 20, 21,
    32, 33, 41–43, 51, 192, 193, 220
raw materials, steelmaking, in, 32–41, 51
reorganization, steel industry, of, in,
    23–32, 50, 92
restrictions, trade, Japanese, on, 7, 48,
    84, 200, 201, 206
Ruhrkohle, coke exports within, 37–38
scrap required in, 40–41
steel-plant construction in, World War II,
    after, 41
subsidies, steel industry, in, 49–51, 230
technology, steel, new, use in, 41–45
Third-World steel markets of, 49, 199
trade, international, steel, and, 47–49
    50, 51, 84, 192, 193, 197, 198, 199
    200, 201, 206, 232, 233, 234
trade, steel, and profits, in, 48
trade, steel, within, 47, 50, 199, 232
transportation, iron ore, of, in, 35–37
transshipment, iron ore of, in, 37
United States: major steel market for,
    48, 51, 116, 206, 231, 234; trade
    agreements, steel, negotiated with, 51
West Germany, coking coal exported to,
    38
European Independent Steel Association, 22
Expansion, steel capacity, of. See Capacity,
    steel, expansion of
Export-Import Bank of Japan, 224
Exports, steel. See Trade, international,
    steel in

Falck group, 30
Fiat SpA, 30
Fifth National People's Conference, China,
    158
Finland, steel industry in, 52, 53, 57
Finsider, Italy, 140
Fire Lake Project. See Sidbec Normines
Firth Sterling, Inc., 113
Florida Steel Corporation, 122
Ford Motor Company, 92, 93, 226
Formed coke, 11, 80, 112
France:
    Arbed, iron ore supplies to, 35
    coal in, 38
    iron ore in, 32, 33, 34, 35
    Minette ore in, 34, 35
    scrap, iron and steel, in, 9, 40, 41,
        45
    Socialist government of, 27, 46

steel industry in, 25, 27–28
    trade, international, steel, in, 6, 193, 205
France, steel industry in:
    basic-oxygen steelmaking in, 42, 45
    capacity in, 27, 28, 221
    closures, plants, by, 27
    continuous casting by, 27, 28
    countervailing duties on, United States,
        by, 205, 212
    Creusot-Loire, 27, 46
    debt, conversion of, government, by, 27
        46
    direct-reduced iron, demand for, in, 45
    EEC, ranking in, 25
    electric-furnace steelmaking in, 28, 42
    employment in, 51
    executive replacement in, 28
    expansion plans, 1974, in, 28
    integrated producers, in, 27
    losses, financial, in, 27, 46
    manhour output in, 27
    Metallurgique de Normandi, 46
    modernization of, 27
    nationalization of, 5, 27, 46, 49
    open-hearth steelmaking in, 42
    ownership of, 5, 27, 46, 49
    production, steel, in, 192, 196
    raw materials for, 32, 33, 34, 35, 38, 40,
        41, 45
    reorganization of, 27–28
    Sacilor, 27, 46
    scrap demand in, 45
    Sollac, 27
    Solmer, 27
    steelmaking processes in, 42, 44
    subsidization of, 5, 46, 49, 205, 216
    Usinor, 27, 46
Fuji Steel, 87
Fundidora Monterrey SA, 166, 167, 168

Georgetown Steel Corporation, 111, 122
Gilmore Steel Company, 202, 237
Granges Steel, 57
Granite City Steel Company, 91
Greece, steel industry in, 49, 52, 200
Greenfield-site plants, steel, 41, 93, 117,
    138, 161, 162
Growth, steel capacity, of. See Capacity,
    steel, expansion of
Growth, steel industry, of:
    boom, steel, and, 15–16
    Canada, in, 137, 222
    capacity, and, 15, 16, 17
    Europe, in, 194, 221
    future, in, xv, 1, 219–223
    Japan, in, 64–67, 68, 76, 88, 194, 222
    1960s and 1970s, in, 2–4, 222
    Non-EEC Western European countries,
        in, 53
    outlook for, steel boom, during, 15, 16

projections for, 4, 221, 222–223
United States, in, 92–96, 119, 194, 221
Gurmendi SA, 171

Hanna Mining Company, 140
Highveld Steel and Vanadium Corp. Ltd.,
    148, 149
Hinduston Steel Company, 164
Hoesch Werke AG (Estel Group), 25, 26,
    31, 40, 43, 51
Hollinger Mines, Ltd., 140
Hoogovens IJmuiden BV, 26, 31, 40, 46, 49
Hungary, steel industry in, 183, 185, 189
HYL direct-reduction process, 167, 168
Hylsa-Hojalata y Lamina SA, 166, 167, 168

IBS-Instituto Brasileiro de Siderurgia, 161,
    231
Imports, steel. See Trade, international,
    steel, in
Inayama, Yoshihiro, 155
Inchon Iron & Steel Company Limited, 165
India, steel industry in, 75, 85, 163, 164,
    189, 194, 195, 213, 215, 216, 229
Indian Iron and Steel Company, 164
Indonesia, steel industry in, 175, 176, 195,
    209
Inland Steel Company, 91, 93, 99, 121, 140
Interlake, Inc., 92, 99, 140
Interministerial Committee for Industrial
    Policy, Italy, 50
International Iron and Steel Institute, 15,
    16, 112, 174
International Steel Committee of OECD,
    214, 215, 235
Interprovincial Steel and Pipe Corp. Ltd.,
    138, 139
Inventories, steel, 17
Investment, steel industry. See Capital
    investment, steel
Iran, 49, 176, 195, 209
Iraq, direct reduction in, 176
Ireland, steel industry in, 32, 40, 42, 46
Irish Steel Holdings, 32, 46
Iron ore. See Raw materials, steelmaking
    for
Iron Ore Company of Canada, 99, 140
Iscor-South African Iron & Steel
    Industrial Corp. Ltd., 148, 149, 150,
    205, 212
Italsider SpA, 21, 29, 30, 49, 50, 140, 162,
    205, 212
Italy:
    Bagnoli, 30,
    coking coal imports of, 8, 38
    Genoa, 30
    government of, 50
    Interministerial Committee for
        Industrial Policy, in, 50
    iron-ore imports of, 34

northern, steel plants in, 30
    scrap imports of, 40
    southern, steel plants in, 29
    steel industry in, 29–30, 40
    Taranto, 29, 30
    trade, international, steel, in, 6, 193
Italy, steel industry in:
    Bagnoli, plant at, 30
    bankruptcy, prospective, in, 30
    basic-oxygen steelmaking in, 42
    Bresciani, 30, 50
    capacity of, 29, 30
    companies in, 29, 30
    composition of, 29
    continuous casting in, 43, 44
    Dalmine, 30
    direct reduction in, 44
    EEC, other steel industries in, and, 30
    efficiency, improved, of, 30
    electric-furnace steelmaking in, 29, 40,
        42, 50
    employment, in, 51
    Falck group, 30, 46
    Fiat, 30, 46
    Genoa, plant at, 30
    Italsider, 21, 29, 30, 46
    Kinglor-Metor process in, 44
    minimills in, 30, 50
    modernization of, 30
    nationalization of, 5, 46
    Nippon Steel, know-how agreement
        with, 30
    open-hearth steelmaking in, 42
    ownership in, 5, 29, 46, 49
    production, steel, of, 56, 196
    raw materials for, 8, 34, 38, 40
    reorganization of, 29–30, 50
    scrap, dependence on, in, 40
    steelmaking processes in, 42
    supsidization of, 5, 49, 50, 205
    Taranto, plant at, 29–30
    Terni, 30
Ivaco, Ltd., 139

Japan:
    automobiles in, 33, 210
    banking system of, steel and, 64–65
    capital-goods demand in, 4
    China: relationship with, outlook for,
        160–161, 209; steel market for,
        159–160, 208–209
    coking coal, imports of, 8, 50, 65–66,
        67, 69–72, 78
    EEC, orderly marketing agreement, steel,
        with, 49, 207
    energy crisis, and, 16, 75
    Export-Import Bank of, Carajas iron-ore
        project, and, 224
    imports, steel, of, 207, 232

iron ore, imports of, 8, 50, 65–66, 67, 68–69, 75, 76
lifestyle, postwar, in, 2
Ministry of International Trade & Industry, 200
shipbuilding in, 3, 75, 82–83
South Korea, and, 169, 209, 232
Taiwan, imports, steel, from, 232
trade, international, steel, in, 2, 48, 66, 81, 82, 84, 192, 193, 194, 197, 198, 199, 200–201, 208–211, 216
United States, automobile exports to, 126, 210
World War II and, 2–3
Japan Iron and Steel Federation, 207
Japan, steel industry in:
annealing lines in, 63, 80
automobiles industry, and, 4, 16–17, 83, 210, 212
basic-oxygen steelmaking in, 68, 73, 76–77, 80
Belgium, steel industry, in, study of, 30
blast-furnace construction in, 63, 65, 66, 68
boom, steel, impact on, 15–18, 63
briquetted coal in, 78, 79–80
bulk-carrier transportation, use of, 37, 74–76, 81–82
capacity: excess in, 67, 208; expansion in, xv, 2, 6, 16–17, 63, 64–65, 67, 68, 88, 222; reduction in, 63, 238
capital investment by, 2, 63, 64, 67, 88, 211, 222
China: coking coal, from 71; export market, steel, for, as, 83, 159–160, 208, 209; steel expansion in, and, 155–159, 161, 213; trade agreement with, 1978, in, 155
coal injection, blast furnaces, for, in, 75
coke production in, 78–80
coke rate in, 80
coking coal, sources of, 70, 71, 72, 80
computer control in, 80
continuous casting in, 63, 64, 66, 77–78
costs in, 2, 63, 64, 75, 80, 88, 210, 227
debt structure in, 64–65
demand, steel, for, and, 63, 67
depression, steel, in, and, 63, 220
diversification by, 88
electric-furnace steelmaking in, 66, 73, 74, 87, 88
energy conservation in, 2, 63, 75, 80
exports, steel: by, 2, 4, 48, 66, 81, 82, 83, 84, 192, 193, 198, 199, 200, 201, 208; cartel, for, 200; indirect, 80, 210; markets for, 83, 84, 85, 116, 160, 194, 198, 200, 201, 208, 209, 231; resistance, overseas, to, 67, 84, 85

Far East, export market for, as, 84, 85, 208, 209
formed coke, pilot plants for, 80
Fuji Steel, 87
Fukuyama, plant at, world's largest, 65
Gilmore Steel Case, decision against, in, 202
growth of, 64–67, 68, 76, 88, 194, 222
Inayama, Yoshihiro, 155
investments, overseas, by, 211, 226
iron ore, sources of, 69, 75, 76
Japan Iron and Steel Federation, 207
Japanese Iron and Steel Company Limited, 87
joint ventures, future, in, 226
Kaiser Steel, pellet contract with, 69
Kawasaki, 63, 64, 87
know-how agreements of, 226
Kobe Steel, 80, 87
loans, banking system to, 64–65
management philosophy of, change in, 221
mergers in, 87
Mitsui & Company, and, 236
modernization of, 2, 63, 88, 210
Muroran Works, Nippon Steel, of, 66
Nippon Kokan, 63, 67, 80, 87, 226
Nippon Steel, 63, 64, 79, 80, 155–156, 210
Nisshin Steel, 87
Ohgishima, Nippon Kokan plant at, 65
oil use in, 75
open-hearth steelmaking in, 68
operating rates in, 2, 85, 208, 222
orderly marketing agreement, EEC, with, 49, 207, 211
outlook for, 66, 67, 72, 86, 87–88
ownership of, 5, 87, 216
pig-iron output in, 68, 70
pollution control in, 64
production, steel, of, 1, 2, 4, 17–18, 64, 67, 68, 76, 81, 82, 83, 87, 192, 196, 208, 220
raw materials for, distant supplies of, 8, 50, 65–66, 67–76, 74–75
scrap, iron and steel, for, 68, 73–74
seacoast location of, 66, 80, 81
seamless-tube mills in, 63–64
size, large, steel plants, of, 65, 80
steel mills, sale of, by, 85, 210–211
Sumi-Coal system of, 79
Sumitomo, 63, 79, 80, 87
Takeda, Yukata, 210
technical assistance by, 85, 88, 155–159, 161, 210–211
technology in, 76–80
Third World: and, 67, 84, 154, 206, 208, 209, 222; competition with, xv,

67, 84–85, 206, 208, 209; export
  market, as, 84, 160, 200, 208; joint
  ventures in, by, 85, 211
trade restrictions on, EEC, in, 7, 48,
  84, 200, 201, 206
United States: coking coal, from, 70,
  72, 101; export market, steel, for, as,
  83, 84, 116, 160, 194, 198, 200, 201,
  208, 209, 231
voluntary restraint agreements, and, 7,
  51, 84, 197–198, 202
West Germany, steel industry, in,
  compared to, 25
Western Europe, export market for, as,
  84, 85, 208, 209
Yawata Steel, 87
Japanese Iron and Steel Company Limited,
  87
Jessop Steel Company, 113
John Lysaght (Australia) Ltd., 146
Joint ventures, steel, 34, 85, 140,
  143, 147–148, 149–150,
  225–226
Jones & Laughlin Steel Corporation, 91,
  99, 107, 113, 121, 140

Kaiser Steel Corporation, 91, 94, 107,
  119, 120
Kawasaki Steel Corporation, 63, 64, 87,
  162
Kennedy, President John F., 117
Kinglor-Metor process, 44
Klockner-Werke AG, 25, 26, 171
KMS process, 26, 43
know-how agreements, steel, 30, 85,
  88, 226
Kobe Steel Limited, 80, 87
Korean War, 92, 106
Korf-Stahl AG, 25, 188
Krupp-Stahl AG, 25, 26, 27

Labrador Mining & Exploration Co. Ltd.,
  140
Laclede Steel Company, 122
Lasco-Lake Ontario Steel Co. Ltd., 138,
  139, 141
LD process. See Basic-oxygen steelmaking
  process
LDAC process, 41–42
LDE process, 43
Liberia, 33, 34, 35, 37, 100
Libya, steel industry in, 175, 176
Lima Declaration on Industrial
  Development and Cooperation, 153
Limestone. See Raw materials, steelmaking
  for
Ling-Tempco-Vought, 113
Lone Star Steel Company, 92, 113, 120

Losses, steel company:
  British Steel Corporation, 24–25
  EEC, in, 1, 18, 19, 20, 21, 49, 50, 96,
    97, 221, 227, 228
  France, in, 27
  Sweden, in, 57, 58
  Third World, in, 178
  United States, in, 91, 96, 97, 127, 201,
    202, 203, 227
  Venezuela, in, 173
  Western Europe, in, 221, 227
  world steel industry, in, 220
Lukens Steel Company, 122–123
Luxembourg:
  coal production of, 39
  coke production of, 39
  France, iron-ore imports from, 33
  iron-ore imports of, 34
  scrap, in, 41
  trade, international, steel, in, 193
Luxembourg, steel industry in:
  Arbed, 28, 35
  basic-oxygen furnaces used in, 29, 42, 45
  blast furnaces, small, replacement of, 29
  continuous casting in, 43, 44
  direct-reduced iron, demand for, in, 45
  electric-furnace steelmaking in, 41
  employment in, 51
  high-phosphorous iron ore, used by, 42
  LDAC process in, 42
  ownership, private, of, 5, 46
  open-hearth replacement in, 29
  production, steel, in, 196
  raw materials for, 33, 34, 39, 41, 42, 45
LWS process, 43
Lykes Steamship Company, 113

Malaysia, steel industry in, 175, 210–211
Manitoba Rollings Mills Ltd., 139
Mannesmann AG, 25, 26, 38, 156
Marshall Plan, 161
Mauritania, 34
MacGregor, Ian, 24
McLouth Steel Corporation, 92, 107, 119
Mergers, steel company:
  Arbed, steel acquisitions by, 28
  Austrian steel industry in, 54
  Cockerill and Sambre, proposed, 30–31
  Davignon plan, under, 23
  Detroit Steel, Cyclops, 91
  EEC, future, in, 51
  Estel, formation of, by, 31
  Failing Company Doctrine, and, 113
  Hoesch and Hoogovens, 31
  Japan, steel industry, in, 87
  Nippon Steel, formation of, by, 87
  Nisshin Steel, formation of, by, 87
  reorganization, steel, and, 219
  Ruhrstahl, formation of, by, 27
  Somisa, Santa Rosa, Gurmendi, and, 171

Svenskt Stål Aktiebolag, formation by, 57
Sweden, steel industry in, 57
Voest-Alpine and Schoeller-Bleckmann, 54
Voest and Alpine, 54
West German steel industry, in, 25
Mexico:
coking coal, lack of, in, 167
direct reduction, in, 11
imports, steel, required in, 167
iron ore, in, 167
OECD Industrial Steel Committee, nonmembership in, 215
steel industry in, 166–168
trade, international, steel, in, 168, 195, 213
Mexico, steel industry in:
Altos Hornos de Mexico, 166, 167, 168
basic-oxygen steelmaking in, 168
blast furnaces of, 167
capacity of, 167
continuous casting in, 167
direct reduction in, 167
electric-furnace steelmaking in, 166, 167, 168
expansion, plans for, 167
Fundidora Monterrey, 166, 167, 168
growth of, 166
HYL direct-reduction process, 167, 168
Hylsa Group, 166, 167
Monterrey plant, 167
open-hearth steelmaking in, 168
outlook for, 168
ownership, public, in, 5, 168, 229
production, steel, of, 166
Puebla plant, 167
Sicartsa, 166, 167, 168
Sidermex Holding Company, 168
Tamsa, 166, 167
Middle East, 49, 194
Midrex direct-reduction process, 141
Minette ore, 34, 35
Minimills:
Australia, steel industry, in, 144
Bresciani, as, 30, 50
Canada, steel industry, in, 137, 138
European Independent Steel Association, and, 22
Italy, steel industry, in, 30, 50
Netherlands, steel industry, in, 31
United Kingdom, steel industry, in, 25
United States, steel industry, in, 93, 94, 109, 119, 122–123, 221
Ministry of International Trade and Industry, Japan, 200–201
Mitsui & Company, 236
Modernization, steel:
Australia, in, 146

Belgian steel industry, requirements of, 31
British Steel Corporation, in, 24–25
China, in, 157, 159
Davignon plan, under, 23–32
EEC steel industry, in, 20, 23ff, 51
French steel industry, in, 27
Hoogovens, plans for, 31
Italian steel industry, in, 30
Japanese steel industry, in, 2, 63, 88, 210
Luxembourg, in, 29
Poland, in, 188
survival strategy, as, 229, 238
Sweden, steel industry, in, 57
United States, steel industry, in, 2, 92, 94, 95, 96, 117–118, 132, 133, 229, 231
West German steel industry, in, 25–27
world, in, 1
Mount Newman Project, 148

National Coal Board, United Kingdom of, 39
National Steel Corporation, 93, 94, 99, 107, 121, 140, 229
Nationalization Act of 1967, 5
Nationalization, steel, 5, 24, 27, 39, 227–228
Netherlands:
coal imported by, 40
countervailing duties, steel, on, U.S., by, 205, 212
iron-ore imports of, 34
scrap in, 41
Netherlands, steel industry in:
basic-oxygen furnaces in, 31, 42, 45
capacity of, 31
continuous casting in, 31, 43, 44
direct-reduced iron, demand for, in, 45
electric furnaces, use in, 31, 42
Estel, 31
Hoogovens, 31, 49, 205
IJmuiden, plant at, 31
integrated producer in, 31
merger, Hoogovens and Hoesch, in, 31
minimill in, 31
modernization plans of, 31
NKF Staahl, 31
ownership of, 5, 46, 47, 49
raw materials for, 34, 40, 41, 45
scrap demand in, 45
seacoast location, benefits of, 31
steelmaking processes in, 42
Thyssen, NFK Staahl subsidiary of, 31
Neunkircher Eisenwerk AG, 29
Nigeria, steel industry in, 174–175, 176, 189
Nippon Steel Corporation, 30, 63, 64, 65, 66, 79, 80, 87, 155, 156, 210, 211

Nippon Kokan KK, 63, 65, 80, 87, 226
Nisshin Steel Company Limited, 87
Nixon, President Richard M., 198
NKF Staahl BV, 31
Non-EEC Western European countries, 48, 49, 51–60
Norrbottens Järnverk AB, 57
North Korea, steel industry in, 5, 174
Northwest Industries, 113
Northwestern Steel & Wire Company, 93, 94
Norway, steel industry in, 52
Nucor Corporation, 122
NVF Company, 113

OBM Process, 43
Obsolescence, steel, in, 1, 49–51, 53, 235
Office of Technology Assessment, 110, 112
Open hearths. See Technology, steel
Operating rates, steel:
    Argentina, in, 170
    Canada, in, 137
    capacity reduction, EEC, in, and, 23, 51
    Japan, in, 2, 85, 208, 222
    reorganization, steel companies of, and, 23, 51
    Sweden, in, 57
    trade problems, and, 237
    Turkey, in, 60
    United States, in, 95, 97, 121, 127, 220, 234
    Venezuela, in, 173
Oregon Steel Mills, 111
Organization of Petroleum Exporting Countries, 16, 175
Ownership, steel companies of:
    Argentina, in, 5, 170, 216
    Australia, in, 5, 145, 216
    Austria, in, 55
    Belgium, in, 5, 30, 46, 49
    Brazil, in, 5, 161, 162, 229
    British Steel Corporation, nationalization of, 5, 24, 45, 228
    Canada, in, 5, 137, 144
    changes in, 1, 2, 4–6, 45–47, 50–51, 227–229
    China, in, 5, 229
    Denmark, in, 46
    Eastern European steel industry, in, 5
    EEC, shifts in, 46–47, 50–51
    France, in, 5, 27, 46, 49
    India, in, 164, 216
    industrialized world, shifts in, 5, 45–47, 50–51, 228
    Ireland, in, 46
    Italy, in, 5, 29, 46, 49
    Japan, in, 5, 87, 216
    Luxembourg, in, 5, 28–29, 46, 49
    Mexico, in, 5, 168, 216, 229

    nationalization, and, 5, 24, 27, 45–47, 228
    Netherlands, in, 5, 46, 49
    1950, in, 4–5
    non-EEC Western European countries, in, 53
    North Korea, in, 5
    private, disadvantages of, 5–6, 227
    problems, with, 5, 49–51, 227–229
    public: employment maintenance, and, 5, 7, 228; excess production, and, 5, 228; losses, financial, and, 5, 7, 138, 143, 144, 227, 228; price effects of, 5, 7, 49, 50, 228; subsidization, and, 5, 7, 49, 50, 51, 138, 143, 219, 228, 229, 230, 231, 235, 236; trade restrictions, and, 7, 50, 219
    public or private, xv, 5, 25, 27, 28, 29, 45, 46, 47, 49–51, 219, 227–229
    South Africa, in, 150
    South Korea, in, 5, 165, 229
    Soviet Union, in, 5
    Spain, in, 56
    Sweden, in, 57
    Taiwan, in, 5, 229
    Third-World, in, 5, 178, 179, 229
    United States, in, 5, 113, 216
    Venezuela, in, 5, 172, 173
    West Germany, in, 5, 25, 45–46, 49
    Western European steel industry, in, 5–6, 24, 25, 27, 28–29, 45–47, 49–51, 55, 227–228

Pakistan, steel-industry projects in, 175, 176
Patent Shaft Steel Works, 25
Penn-Dixie Cement, 113
Penn-Dixie Steel Corporation, 122
Peru, steel industry in, 75, 176
Philippines, steel industry in, 85, 175, 209
Phoenix Steel Corporation, 122–123
Picands-Mather & Company, 140
Pohang Iron and Steel Company Limited, 165
Poland, steel industry in, 9, 38, 39, 40, 48, 55, 184–185, 186, 187, 188, 224
Pollution control, steel industry, in, 2, 45, 64, 97
Portugal, steel industry in, 52–53
Prices, steel:
    boom, steel, impact on, 16
    concessions, trade, and, 6–7, 48, 49, 50, 194, 199, 201, 204, 205, 228
    cutting of, EEC, in, 22, 48, 49, 50
    Davignon plan, under, 20
    EEC, in, 1, 7, 19, 20, 22, 49, 50
    government ownership, and, 5, 7
    increase in, Eurofer II, under, 23
    monitoring, EEC in, Eurofer II, under, 22

reductions, in, 5
stability of, Eurofer II, under, 23
structure of, United States, in, 5, 97, 128,
    199, 201, 202, 203, 204, 205
trade, impact of, on, 5–7, 49–50, 97, 116,
    128–129, 130, 131, 228
trade restrictions, as, EEC, in, 48, 207
trigger-price mechanism, 5, 7, 48,
    97, 116, 129–130, 203, 204, 207,
    230, 236
United States, imports of, and, 128–129,
    130, 131
Profits. *See* Earnings, steel
Propulsara Siderurgica SA, 170

Quebec-Cartier Mines, 99, 139, 140
Quotas, import, steel:
Davignon plan, under, 20
EEC, in, 7, 20, 48, 84, 200, 201, 206, 207
voluntary, United States, in, 7, 51, 84,
    197–198, 202, 237

Raritan River Steel Company, 122
Rationalization, steel. *See* Reorganization,
    steel companies of
Raw materials, steelmaking for:
Australia, in, 144–148
Austria, for, 53, 55
availability of, xv, 7–8, 15, 32–37,
    223–224
Canada, in, 139–140
charcoal, use of, Brazil, in, 162
China, in, 157
coke: Arbed, coke batteries of, 38;
    British Steel, production of, 39; coal
    blending, production of, in, 39; Coal
    Products Limited, production of, 39;
    Eastern Europe, imports of, United
    Kingdom, from, 186; EBV, subsidiary,
    Arbed of, 39; Estel, Hoogovens
    division of, coke ovens of, 40;
    Mannesman, coke batteries of, 38;
    Ruhrkohle, capacity of, 38; Salzgitter,
    coke batteries of, 38; technology,
    production, for, 9, 79–80; Thyssen,
    coke batteries of, 38; United States,
    shortages of, prospects for, 102–103
coking coal: blending of, 39, 40;
    briquetting of, 9, 78, 79–80; British
    Steel, for, 39; Coal Arbed, mines of,
    United States, in, 39; direct reduction,
    and, 10; discoveries of, 9; European
    steel industry, imports of, dependence
    on, 40; exports of, 38, 39, 40, 55, 70,
    71, 72, 80, 101–102, 146, 167; imports
    of, 8, 37–40, 55, 59, 65–66, 67, 69–72,
    78, 80, 146; inferior, use of, 9, 39, 40;
    metallurgical, 7, 8–9, 37–40, 224;
    National Coal Board, Coal Products
    subsidiary of, 39; prices of, steel boom,

and, 9; production, reserves, and, 9,
    37, 38, 39, 101–103, 146, 150, 186,
    224; Ruhrkohle, West Germany, and,
    38; Sidmar, joint-venture mine, United
    States, in, 38; subsidies for, 39
depletion of, 7, 32, 98
direct-reduced iron, 11, 44–45, 56, 103,
    110, 140, 141, 167, 171, 172, 174–175,
    176, 179, 188, 209, 224, 225
EEC, in, 32–41, 51
Fire Lake Project, 99
Hollinger Mines Ltd., 140
India, in, 164
iron ore: beneficiation of, 8, 35, 37, 98,
    99, 101, 112; capital investment in, 8;
    Carajas Project, Brazil, in, 223–224;
    competition in, 148; deposits, location
    of, United States, in, 97, 98, 99, 101;
    discoveries of, 8, 98; distant supplies
    of, 8, 32–37; exports of, 33, 34, 35,
    37, 51, 55, 60, 69, 75, 76, 98–99, 100,
    146–147; imports of, 33, 34, 35, 36,
    37, 51, 55, 65–66, 69–72, 75, 76, 78,
    98–101, 146, 186; Iron Ore Company
    of Canada, 99; joint ventures in, 34;
    Kaiser Steel, pellets supplied, Japan to,
    69; Labrador Mining & Exploration
    Co. Ltd., 140; low-grade, 8, 33, 34,
    37, 60, 101; Minette, France, in, 34,
    35; pellets, 8, 35, 69, 98; Port Cartier,
    140, 143; price of, 8; production
    reserves, and, 8, 32, 33, 34, 35, 37,
    60, 98, 99, 100–101, 139–140, 146, 148,
    150, 161, 163, 172, 185, 223–224;
    Quebec-Cartier Mines, 99; Sidbec
    Normines, 140, 143; taconite, 101;
    trade in, 8, 32–37; transportation of,
    34, 35–37; Wabush Mines, 99
Japan: costs of, rising, in, 75; distant
    supplies of, for, 8, 50, 65–66, 67–76,
    74–75
limestone, 7
need for, 1, 45, 223
non-EEC Western European countries,
    in, 52
post-World War II period, in, 7–10,
    32–37
scrap, iron and steel: continuous casting,
    availability of, and, 9, 41; demand for,
    7, 9, 32, 40, 41, 45; EEC, steel
    industries, electric furnace, and, 7, 9,
    32, 40, 41; EEC, steel industries,
    exports of, 9, 40; imports of, 40;
    mill-revert, 9, 41, 103–104; quality,
    deterioration of, 9, 10, 41, 224;
    residual elements in, 9, 104, 122, 123;
    sources of, 103–104; supply future, of,
    9, 40, 224; supplies of, 9, 41, 44, 45,
    103–105; United States, requirements
    for, 103

sinter, 8
sources of, shifts in, 2, 32–41, 51, 98,
    223
South Korea, bulk-carrier transportation,
    and, 165
supply of, limitations on, 15
Taiwan, imports of, 169
Third World, in, 178
transportation of, costs of: fuel-oil prices,
    and, 37; future increases in, 37
United States, in, 8, 9, 97–105
Western Europe: distant supplies of, for,
    8, 32–41, 51; prewar supply of, 32
Reorganization, steel companies of:
    bankruptcy, and, 219
    Belgium, in, 30–31
    capacity reduction, and, 23, 51, 219, 221
    Davignon plan, under, 20, 23ff
    Denmark, in, 32
    diversification by, 221
    EEC, in, 23ff
    factors contributing to, 219
    France, in, 27–28
    future, in, xv, 219
    Ireland, in, 32
    Italy, in, 29–30
    Japan, in, 221
    Luxembourg, in, 28–29
    management philosophy, and, 220–221
    mergers, through, 23, 50–51, 91, 113,
        132, 219
    modernization, through, 23
    Netherlands, in, 31
    operating rates, and, 23, 51
    profit objectives of, 23, 221
    subsidies, government, and, 50, 51, 219
    Sweden, steel industry in, 57
    United Kingdom, in, 24–25, 221
    United States, in, 2, 91, 92, 113, 132,
        133, 221
    West Germany, in, 25–27
Republic Steel Corporation, 91, 94, 99,
    110, 121, 140
Restructuring, steel. See Reorganization,
    steel companies of
Roderick, David M., 234
Romania, exports, steel, EEC, to, 48
Round Oaks Steel Works, 25
Rouge Steel Company, 92, 226
Ruhrkohle-Verkauf GMBH, 37–38
Rumania, steel industry in, 183, 185, 186,
    187, 188–189

Sacilor, Acieries et Laminoris de Lorraine,
    27, 46
Salzgitter. See Stahlwerke Peine-Salzgitter
Sambre, Ste Metallurgique
    Hainaut-Sambre SA, 30, 31, 46, 205
Sandvik AB, 58
Saudi Arabia, 49, 175, 176

Scaw Metals Ltd, 148
Schoeller-Bleckmann Stahlwerke AG, 54
Scrap, iron and steel. See Raw Materials,
    steelmaking for
Sevensa-Siderurgica Venezolana SA, 173
Sharon Steel Corporation, 91, 92, 113
Sheerness Steel Company, Ltd., 25
Sherman Anti-Trust Act, 198–199
Shipbuilding, 3, 4, 59, 75, 82–83
Shortages, steel, 16
Sicartsa-Siderurgica Lazaro Cardenas-Las
    Truchas SA, 166, 167, 168
Sidbec-Dosco, Ltd., 137, 138, 139, 140,
    141, 143
Sidbec-Normines, 140, 143
Siderbras-Siderurgia Brasileira SA, 162
Sidermex Holding Company, 168
Sidmar, Siderurgie Maritime SA, 28, 30, 31,
    38, 46
Sidor-CVG Siderurgica del Orinoco CA,
    172, 173, 174
Simonnet plan, 20
Sishen-Soldana Bay Project, 149
Size, steel companies of:
    Austria, in, 51, 54
    British Steel Corporation, reduction of,
        in, 25
    Cockerill, reduction of, proposed, 31
    efficiency, and, 51
    Japan, in, 65, 80
    Neunkircher, reduction of, in, 29
    operating rates, and, 51
    reduction in, 23, 31, 51, 238
    Spain, in, future, 51, 56
    survival and, xv, 238
SKF Steel Division, 58
Slater Steel Industries, Ltd., 139
SLRN direct-reduction process, 141
Solmer, Ste Lorraine et Meridionale de
    Laminage Contina, 21, 27, 28
Somisa-Sdad Mixta Siderurgia Argentina,
    170, 171
South Africa:
    coal exports of, 38, 40, 150
    countervailing duties, steel, on, U.S.,
        by, 205, 212
    exports, steel, of, 85
    iron ore in, 150, 224
    iron-ore supplies, EEC, to, 35
South Africa, steel industry in:
    basic-oxygen steelmaking in, 149, 150
    blast furnaces of, 149
    briquetting, coal, in, 149
    capacity of, 148, 149, 222
    capital investment projects, in 1973,
        149
    companies in, 148
    continuous casting in, 150
    direct reduction in, 149
    Dunswart Iron and Steel, 148, 149

electric furnaces in, 148, 149, 150
growth of, 148, 222
Highveld Steel, 148, 149
Iscor, 148, 149, 205
joint-venture steel plant, proposed,
    149–150
Newcastle Works, 149
open-hearth steelmaking in, 150
ownership of, 150
Pretoria Works, 148
production, steel of, 148
Scaw Metals, 148
Sishen-Saldana Bay project, 149
Southern Cross Steel, 148
technology in, 150
trade, international, steel, in, 150
Vanderbijlpark Works, 149
South America, iron-ore supplies, EEC,
    to, 33, 34, 37, 51
South Korea:
    coal, coking, lack of, in, 8
    exports, steel, of, 85, 166, 232
    Japan, and, 84, 166, 208, 209, 232
    OECD International Steel Committee,
        nonmembership in, 215
    steel demand, in, 177
    steel industry in, 165–166
South Korea, steel industry in:
    basic-oxygen steelmaking in, 165, 166
    blast furnaces in, 165
    bulk-cargo carriers, and, 165
    capacity of, 165
    coal, coking, lack of, 8
    continuous casting in, 165, 166
    Dongkuk Steel Mill Company, 165
    electric-furnace steelmaking in, 165, 166
    expansion plans, steel, of, xv, 165, 222
    Far Eastern markets, steel, and, 166
    growth of, 165
    Inshon Iron and Steel Company, 165
    Japan, and, 84, 166, 208, 209, 232
    Kisco, 165
    labor costs in, 232
    outlook for, 166, 222
    ownership, in, 5, 165, 229
    Pohang Steel Company, 165
    production, steel, in, 165
    trade, international, steel, in, 84, 166,
        177, 194, 195, 208, 209, 213, 215, 232
Southern Cross Steel Company Ltd., 148
Soviet Union, steel industry in, xv, 5, 25,
    38, 48, 55, 154, 160, 183–189, 192, 200,
    206, 227
Spain:
    agreement, trade, EEC, Japan, on, 49
    competition, lack of, in, 235
    EEC peripheral steel market, as, 49
    exports, steel, of, 48, 85
    imports, steel, Japan, from, 49, 200
    leading non-EEC steel producer, as, 55

    obsolescence in, 235
    steel production in, 52, 55, 56
Spain, steel industry in:
    Altos Hornos de Vizcaya, 56
    Altos Hornos del Mediterraneo, 56
    Aviles, plant of, 56
    basic-oxygen furnaces in, 56
    capacity of, 56
    continuous casting, additional, for, 56
    direct reduction, in, 56
    electric-furnace companies in, 56
    Ensidesa, 56
    exports, steel, EEC, to, 48
    finishing facilities in, 56
    government role in, 56
    growth, rapid, of, 49, 55
    integrated producers in, 56
    market, domestic, of, 57
    open-hearth, replacement of, 56
    ownership of, 56
    production of, 52, 55, 56
    regrouping in, 56
    size of, future, in, 51, 56
    technology, used in, 56
    Uninsa, acquisition, Ensidesa, by, 56
    Verina, plant at, 56
Stahlwerke Peine-Salzgitter AG, 26, 27,
    38, 45, 46
Stahlwerke Rochling-Burbach GmbH, 25
Stahlwerke Sudwestfalen AG, 26
Ste des Acieries et Trefileries de
    Neuves-Maisons Chatillon SA, 46
Ste Metallurgique de Normandie, 46
Steel Authority of India Limited, 164
Steel demand. See Demand, steel
Stelco, Inc., 137, 138, 140, 141
Stora Kopparberg Bergslags AB, 57
Subsidies, steel:
    Arbed, West German government grant
        to, 29
    Belgian steel industry, in, 5, 49
    Canada, steel industry, in, 138, 143
    coking coal, for, 39, 224
    controversy, steel industry, EEC, in,
        and, 49–50
    dependence on, 1
    disadvantages, private companies, to,
        5–6, 49–50, 229
    employment maintenance, and, 228
    French steel industry, in, 5, 49
    future of, 227, 228, 236, 238
    government ownership, and, 5, 49, 50,
        51, 219, 228, 230, 236
    Interministerial Committee for Industrial
        Policy, Italy, and, 50
    Italian steel industry, in, 5, 49, 50
    losses, and, 5, 7, 49, 50, 51, 138,
        143, 228
    price effects of, 5, 7, 49, 50, 128–129,
        130, 131, 228

reorganization, steel industry, and, 50, 219
Sweden, steel industry, in, 58
Third World, in, 177–178
trade, restrictions on, and, 7, 50
United Kingdom steel industry, in, 25, 39, 49, 224
United States, steel imports of, and, 5, 97, 116–117, 128, 129, 130, 131, 133, 191, 205, 211, 214, 231, 234
West German steel industry, in, 27, 29, 37, 50, 224, 229
Western Europe, steel industry, trend toward, in, 227–228
withdrawal of, 235
Sumi-Coal system, 79
Sumitomo Metal Industries Limited, 63, 79, 80, 87
Svenskt Stål Aktiebolag (SSAB), 57, 58
Sweden:
coal imports, United States, from, 59
exports, steel, EEC, to, 48
iron ore in, 32, 33, 34, 35, 57, 59
technology, steel, in, 59
trade, international, steel, in, 59
Sweden, steel industry in:
basic-oxygen steelmaking in, 53, 59
capacity of, 58
continuous casting in, 58, 59
Domnarvet, plant at, 57, 58
efficiency in, 59
electric-furnace operations in, 57
expansion plans, cancellation of, 58
government participation in, 57
Granges, 57
loans, government, to, 58
losses, financial, of, 57, 58
Luleå, plant at, 57, 58
market objectives of, 57
Norrbottens, 57
operating rates in, 57
outlook for, 59
ownership of, 57
Oxelösund, plant at, 57, 58
production by, 52, 58
reorganization of, 57
Sandvik AB, 58
size of, future, in, 51
SKF Stål, 58
specialty-steel producers in, 58
Stora-Kopparberg, 57
survival of, future, in, 57
Svenskt Stål Aktiebolag, formation of, 57
Switzerland, steel industry in, 52, 53
Sydney Steel Corporation, 137, 139, 141, 143, 227

Taiwan:
coking coal, imports of, 169

exports, steel, of, 85, 232
iron ore, imports of, 169
steel industry in, 168–170
Taiwan, steel industry in:
basic-oxygen steelmaking in, 169
blast furnaces of, 169
capacity of, 85, 169, 222
China Steel, 168, 169
continuous casting in, 169
electric-furnace producers in, 169
finishing facilities of, 85, 169
growth of, 168
Japan, trade, and, 85, 169, 208, 209, 232
ownership, public, of, 5, 229
production, steel, of, 168
raw materials for, 169
trade, international, steel, in, 169, 170, 177, 208, 209, 213, 232
U.S. Steel, technical assistance from, 85
Taketa, Yukata, 210
Tamsa-Tubos de Acero de Mexico SA, 166, 167, 168
Technology, steel:
AOD process, 45
argon oxygen decarburization, 12
basic-oxygen steelmaking, 9, 10, 26, 27, 28, 29, 30, 31, 41–43, 45, 51, 52, 53, 54, 55, 59, 68, 73, 76–77, 80, 96, 106, 107–108, 122, 138, 140, 141, 146, 147, 149, 150, 162, 164, 165, 166, 168, 169, 170, 172, 184, 185, 188, 224
Bessemer converters, replacement of, 27, 29
blast furnaces, 7, 8, 9, 21, 24, 26, 27, 41, 42, 45, 51, 54, 63, 65, 66, 68, 91, 93, 94, 95, 98, 101, 102, 138, 145, 147, 156, 157, 165, 167, 169, 170, 185
briquetted coal, 78, 79, 80, 149
capital costs, and, 12, 94–95, 101, 103, 131, 224–227
coke ovens, 38, 39, 45, 78–80, 102–103, 185
competitiveness, steel, and, 12, 231
computers, application of, 10, 12, 45, 80, 112
continuous annealing, 12, 63, 80, 113
continuous casting, 10, 11, 27, 28, 29, 30, 31, 32, 41, 43–44, 52, 53, 58, 59, 63, 64, 66, 77–78, 94, 108–110, 122, 124, 141, 146, 147, 150, 162, 165, 166, 167, 168, 169, 171, 172, 184, 185, 221, 224, 225, 230
direct reduction, 10, 11, 44–45, 103, 110–112, 167, 168, 171, 172, 174, 175, 176, 179, 188
dry quenching, coke, of, 185
electric furnaces, 7, 8, 9, 21, 24, 26, 29, 30, 31, 32, 40, 41, 42, 43, 45, 52, 53, 54, 57, 66, 73, 74, 88, 91, 96, 103, 105, 108, 122, 123, 140, 141, 147, 148,

149, 150, 162, 164, 166, 168, 169, 171,
  172, 174, 184, 185, 221, 224, 225
energy saving, and, 10
formed coke, 11, 80, 112
iron desulfurization, 12
iron-ore beneficiation, 12, 98, 99, 101,
  112
Kaldo converters, replacement of, 27, 29
Kinglor-Metor process, 44
LDAC process, 41, 42
LDE process, 43
LWS process, 43
minimill, 22, 25, 30, 31, 50, 93, 94, 109,
  119, 122–123, 137, 138, 144, 188
new, application of, 2, 41–45, 51, 55
new, need for, 1
non-EEC Western European countries,
  in, 52, 56
OBM process, 43
open hearth, replacement of, 10, 26,
  28, 29, 30, 42, 43, 51, 54, 68, 94,
  105–107, 121–122, 140, 141, 143, 145,
  146, 147, 150, 162, 164, 168, 172, 174,
  184, 185, 188, 224, 235
pipeline charging, coke ovens, for, 45
pollution control, 2, 45, 64, 97
post-World War II period, in, 10–12,
  41–45
refractories, blast furnace, 45
Sumi-coal system, 79
temperatures, blast furnace, 45
tops, blast furnace, 45
ultrahigh power, electric furnaces, in, 12,
  112, 122, 224
use of, xv, 41–45
water-cooled panels, electric furnaces,
  in, 112, 122, 225
yield improvement, and, 10, 11
Teledyne, 113
Terni SpA, 30
Third World:
  Algeria, steel industry in, 175
  Bangladesh, steel plans in, 175
  boom, steel, impact on, 18
  Brazil, steel industry, in, 126, 155,
    161–163, 176, 177, 178, 179, 195, 213,
    215, 229, 231, 232
  capacity in, 2, 154, 176, 177, 206, 213
  Chile, steel industry, in, 174, 176
  China, steel industry, in, 154–161, 206,
    209, 229
  Columbia, steel industry, in, 176
  cost competitiveness, steel, in, 177–178
  costs, new technology of, in, 227
  direct reduction, use of, in, 176, 179, 225
  economic diversity in, 175–176
  Egypt, steel industry, in, 175, 176
  expansion program, steel, in, xv, 2, 153,
    175, 176, 222
  India, steel industry, in,

Indonesia, steel industry, in, 175, 176,
  195, 209
industrialized world, steel industry in,
  impact on, 177
Iran, in, 176, 195
iron-ore trade in, 191, 224
Japanese competition with, xv, 67, 84–85,
  206, 208, 209
Japanese steel industry, and, 67, 84, 154,
  160, 200, 208, 222
labor costs, steel in, 178
Libya, steel industry, in, 175, 176, 178
Lima Declaration, steel, and, 153
losses, financial, steel, in, 178
Malaysia, steel industry, in, 175, 210–211
Mexico, steel industry, in, 155, 166–168,
  178, 179, 195, 213, 215, 216, 229
modern equipment, steel, in, 178–179
Nigeria, steel industry, in, 174–175, 176,
  178, 179
North Korea, steel industry, in, 155, 174
OPEC countries in, 175
outlook, growth, steel, in, for, 175–179
ownership, public, steel, in, 5, 178, 179,
  216, 229
Pakistan, steel plans in, 175, 176
Peru, steel industry, in, 176
Philippines, steel industry, in, 175
plans beyond 1985, xv, 153, 177
production, steel, by, 2, 153–154, 213
raw materials, steelmaking, in, 178
subsidies, steel, in, 177–178
Taiwan, steel industry, in, 154, 155,
  168–170, 177, 178, 179, 209, 213, 229
technology, steel, in, 178–179
trade, international, steel, and, 154, 194,
  195, 200, 206, 207, 213, 216, 222
trade restrictions in, 207
United Nations Industrial Organization,
  steel expansion, and, 153
United States, steel market in, and, 231
Venezuela, steel industry, in, 155,
  172–174, 178, 179, 195
Thyssen Aktiengesellschaft, 25–26, 38, 43,
  45, 157, 229
Trade, international, steel, in:
  boom, steel, impact on, 16, 18, 199, 200,
    207, 210, 230
  Canada, in, 141, 142, 143, 198
  closings, plant, and, 201–202
  crisis in, 1977, in, 201–204
  difficulties in, 1, 219, 228, 230–238
  EEC: Davignon Plan, and, 20; export
    markets of, 48, 49, 51, 116, 199, 206
    231; exports by, 6, 47–48, 51, 192–193,
    198, 199, 201; imports of, 47–48, 55,
    200, 201; in, 47–49, 50, 51, 192, 193,
    197, 198, 199, 200, 201, 206, 232;
    orderly marketing agreement, Japan,
    with, 49, 207; peripheral markets of,

Japan, and, 49, 200, 207, 211; price
  cutting in, and, 48, 49, 50; profits,
  in, and, 48, 199; quotas on, in, 7, 20,
  48, 206, 207; restrictions on, in, 7, 20,
  48, 50, 84, 200, 201, 206, 207
employment maintenance, and, 2, 5, 7,
  191, 204, 206, 228
future of, xv, 206, 219, 230–232, 237
growth of, 2, 6, 191
Japan: capacity, steel, in, and, 6, 191,
  194, 208; export markets of, 83, 84,
  116, 159, 160, 194, 198, 200, 201, 208,
  209, 210, 231; exports by, 2, 48, 66,
  81, 82, 83, 84, 192, 193, 198, 199, 200,
  201, 208; Gilmore Steel dumping case,
  and, 202; imports of, South Korea,
  from, 166, 207, 232; in, 81–87, 191,
  192, 193, 194, 197, 198, 199, 200–201,
  208–211, 216; indirect exports of,
  United States, to, 212; leading position
  of, 6, 191, 206; Ministry of
  International Trade, export cartel of,
  200–201; Mitsui & Company, trigger
  prices circumvented by, 236; orderly
  marketing agreement, EEC, with, 49,
  207; peripheral markets of, EEC, and,
  49, 200, 207, 211; resistance to,
  exports of, 67, 84, 85, 197, 200–201;
  restrictions on, EEC, in, 7, 48, 84, 200,
  201, 206; Taiwan, with, 169, 208, 209,
  232
newly industrialized countries, exports of,
  85, 86
non-EEC Western European countries,
  and, 48, 49, 52–53, 55, 56–57, 193, 200
ownership, steel, and, 191, 230
price concessions, role of, in, 6–7, 48, 49,
  50, 130, 131, 194, 199, 201, 204, 205,
  214, 228
production, steel, and, 191, 192
protectionism, and, 191, 197, 200,
  206–208
restrictions on, 7, 18, 48, 50, 51, 191,
  197, 200, 201, 203, 205, 214, 215,
  216, 230
solutions, proposed, problems, to,
  233–238
Soviet Bloc, in, 160, 186–187, 189, 200,
  206, 210
tariffs, temporary, and, 7, 197
Third World, in, 49, 85, 154, 162, 163,
  164, 166, 168, 169, 170, 172, 173, 177,
  191, 194, 195, 199, 200, 205, 206, 208,
  209, 213, 215, 216, 231, 232
United States: American Iron and Steel
  Institute, and, 197; Antidumping Act
  of, 203, 207, 236; antidumping suits
  filed in, 116, 118, 130, 202, 203, 204,
  207; Baldrige, Commerce Secretary,
  and, 205; bankruptcies, steel, and, 7,
  201; Brock, Trade Representative, and,
  205; Carter, President, and, 203, 204;
  Commerce Department, and, 204, 205;
  Consumers Union, and, 198;
  countervailing duties of, 205, 207, 212,
  233, 234; employment, steel, and, 7,
  202, 234; export target, major, EEC,
  for, 48, 51, 116, 206, 231; exports by,
  115, 192–193, 197, 199; impact on,
  6–7, 91, 97, 113–119, 127–132, 191,
  194, 197, 201–202, 234; imports,
  dumping of, in, 97, 116, 129, 130, 131,
  202, 203, 205, 212; imports, exchange
  rates, and, 130, 131, 204; imports,
  factors increasing, 114, 191, 194;
  imports, labor negotiations, and,
  114–115, 194; imports, need for,
  127–128, 194; imports of, 7, 141, 191,
  194, 197, 198, 200, 201–206, 211–212,
  233, 234, 236; imports, prices of,
  128–129, 130, 131; imports, problem
  of, in, 127–132, 197, 201–206;
  in, 5, 6, 7, 91, 97, 113–119, 127–132,
  191, 192, 193, 194–200, 201–206, 233
  International Trade Commission of,
  and, 204, 205; Nixon administration,
  import surcharge of, 198;
  preclearance, trigger prices, and, 204;
  price structure, steel, in, and, 5, 7, 128,
  130, 131, 199, 201, 202, 203, 204, 205;
  profits, steel, and, 7, 201, 202, 203;
  trigger-price mechanism, and, 5, 7, 48,
  97, 116, 129–130, 203, 204, 205, 207,
  230, 236; United States Steel,
  antidumping suits of, 203, 207, 236;
  United Steelworkers of America, and,
  197
voluntary restraint agreements, and, 7,
  51, 84, 197–198, 202, 237
wage rates, low, and, 194
Western European capacity, and, 6,
  191, 194, 211
World War II, changes in since, 191
Transportation, raw materials, steelmaking
  for, 35–37, 74–76, 81–82
Trigger-price mechanism:
  Antidumping Act, U.S., and, 203, 236,
    237
  antidumping investigations, and, 203
  Canadian steel industry, and, 142
  Carter steel program, in, 203
  circumventing of, 205, 233, 236
  Commerce Department, U.S., and, 204,
    233, 237
  countermeasure to subsidies, as, 5
  dumping investigations under, 204
  imports, steel, U.S., reduction of, and,
    203
  improved version of, 204
  preclearance to avoid, 204

stabilizing steel prices, and, 5, 203, 204
steel import prices, and, 5, 203, 204
suspension of, 205, 207
U.S. government, adoption of, 5, 203, 207
United States, steel industry in, and, 5, 7, 48, 97, 116, 129–130, 203, 204, 205, 207, 230, 236
Truman, President Harry S., 106
Turkey, steel industry in, 60
Tysland-Hole process, 172

United Kingdom:
British Trade Secretary, 234
coke exported by, 39
coke imported by, 39
coal producer, as, 39
Coal Products Limited in, 39
coal subsidies in, 39, 224
coal-mine closings in, 39
countervailing duties on, U.S., by, 205, 212, 234
direct reduction in, 44
iron ore produced in, 33–34
iron-ore imports of, 34
National Coal Board in, 39
nationalization, steel industry, of, 5, 24, 45
scrap, dependence on, in, 40
scrap exported by, 9
scrap exports by, 40
steel industry in, 24–25, 45
steelmaking processes in, 42
voluntary import restraints, U.S., nonparticipation in, 198
United Kingdom, steel industry in:
Alphasteel, 25
basic-oxygen steelmaking in, 42
British Steel Corporation, 5, 21, 24–25, 45
capacity of, 24, 45
countervailing duties on, U.S., by, 205, 212
direct reduction in, 44
Duport Steel Limited, closing of, 25
electric-furnace steelmaking in, 40, 42
exports, steel, by, 192, 193
nationalization of, 5, 24, 45
ownership of, 5, 24, 25, 45, 49
Patent Shaft Steel Works, closing of, 25
private sector of, 24, 25, 45, 49
production, steel, in, 192
raw materials for, 33, 34, 39, 40
reorganization of, 24–25, 45
Round Oaks Steel Works, closing of, 25
scrap, dependence on, in, 40
Sheerness Steel Company, Ltd., 25
subsidies, government, of, 49, 205
United Nations Industrial Organization, 153

United States:
Alabama, iron ore, in, 101
Antidumping Act of, 203, 207
antidumping suits filed in, 116, 118, 130, 202, 203, 204, 207, 236, 237
automobile imports in, 125–126
automobile industry in, 3, 17–18, 124–127
automobile production in, 124–125
automobile treaty, Canada, with, 142
California, iron ore, in, 101
Canada, steel imports, from, 141
capital-goods demand in, 4
Carter, President Jimmy, of, 203, 204
Coal Arbed, mines of, in, 39, 102
coal, coking, in, 9, 101, 224
coal exports of, 38, 39, 40, 101–102, 140, 143
Commerce Department of, 204, 205, 233, 234, 237
Consumers Union of the United States, 198
countervailing duties, steel, in, 205, 207, 212, 233, 234
dumping, imports, steel, in, 97, 116, 129, 130, 131, 202, 203, 205
EEC steel-export market, as, 48, 51, 116
EEC, trade agreement, negotiated, with, 51
export market, steel, Japan, for, 83, 84, 116, 160
exporter, steel, as, 113
importer, steel, as, xv, 113
imports, steel: exchange rates, and, 130, 131, 204; factors increasing, 114, 191, 194; into, 5, 6, 7, 91, 113–119, 127–132, 191, 192, 193, 194–200, 201–206; labor negotiations, and, 114–115, 194; need for, 127–128, 194; prices of, 128–129, 130, 131, 194, 199, 201, 204, 205
indirect steel imports, Japan from, 212
International Trade Commission of, 204, 205
iron ore in, 8, 98, 101
Japan, automobile imports from, 126
Japan, raw materials shipped to, 73, 74, 75
Kentucky, coal reserves, in, 101
New York, iron ore, in, 101
Nixon administration, import surcharge of, 198
Pennsylvania, eastern, iron ore, in, 101
price structure, steel, in, 5, 7, 128, 199, 201, 202, 203, 204, 205
scrap in, 9, 40, 103
scrap exports of, 104–105
Sidmar, joint-venture coal mine of, in, 38
steel industry in, 91–135
steel-import problem in, 127–132, 197, 201–206

subsidized imports, steel, in, 5, 97, 116–117, 128, 129, 130, 131, 133, 191, 203, 205, 211, 214
Sweden, coal exports to, 59
trade, international, steel, in, 5, 6, 7, 91, 97, 113–119, 127–132, 191, 192, 193, 194–200, 201, 206, 231, 232, 233, 234, 236, 237, 238
trigger-price mechanism, steel, in, 5, 7, 48, 97, 116, 129–130, 203, 204, 205, 207
Truman, President Harry S., of, 106
West Virginia, coal reserves, in, 101
Wyoming, formed coke production, in, 11
Wyoming, iron ore, in, 101
United States Congress, 7, 237
United States Customs Service, 236
United States Department of Commerce, 204, 205, 212
United States Department of the Treasury, 203
United States International Trade Commission, 204, 205
United States government:
    Consumers Union suit against, 198–199
    trigger-price mechanism, adoption of, by, 5, 203, 207
United States Senate:
    hearing of, 197, 237
    quota bill, imports on, 197
United States Steel Corporation, 85, 91, 93, 94, 95, 99, 109, 114, 117, 119, 120, 121, 124, 139, 140, 143, 157, 173, 203, 207, 227, 229, 234
United States, steel industry in:
    Alan Wood Steel, 91, 92, 132, 133
    American Iron and Steel Institute, 96, 197
    antidumping suits filed by, 116, 118, 130, 202, 203, 204, 207
    Armco Steel, 91, 95, 99, 110, 111, 117, 123, 227
    Atlantic Steel, 122
    automobile industry and, 3, 17–18, 124–127, 212
    bankruptcies in, 7, 18, 91, 93, 96, 113, 201
    basic-oxygen steelmaking in, 68, 96, 106, 107–108, 122
    Bayou Steel, 122
    beneficiation, iron ore, by, 98, 99, 101, 112
    Bethlehem Steel, 91, 93, 94, 95, 96, 97, 99, 117, 119, 120, 121, 124, 202, 227
    Birmingham, Alabama, in, 120
    blast furnaces of, 91, 93, 94, 95, 98, 101, 102
    boom, steel, impact on, 15–18, 92, 128, 129
    capacity of, 92, 93, 95, 96, 105–106, 117, 119, 120, 122, 123, 126, 127, 129, 131, 132, 133, 202, 212, 221, 229
    capital costs in, 94–95, 101, 103, 131
    CF&I Steel, 92, 94, 113
    coal, coking, supply of, 9, 97–98, 101, 103, 224
    closings, plant, in, 91, 93, 97, 117, 119, 120, 121, 130, 201–202
    coastal locations, and, 120–121
    coke-oven capacity of, 102–103
    coke, shortages, prospects for, 102–103
    Colt Industries, 113
    Columbia Steel, 113
    computer control in, 112
    conglomerate take-overs, and, 113
    Continental Steel, 113
    continuous annealing in, 113
    continuous casting in, 94, 108–110, 122, 124, 212, 225
    cost comparisons, in, 131
    Crane, 113
    Crucible Steel, 91, 93, 113
    cyclical fluctuations in, 4
    Cyclops, 91, 93
    demand, steel, for, and, 95, 97, 105–106, 114, 127–128, 132
    Detroit Steel, 91
    direct reduction, in, 11, 103, 110–112
    direct-reduced iron, production of, 103, 110
    diversification in, 2, 133, 221, 229
    dumping, imports, steel, in, 97, 116, 129, 130, 131
    earnings in, 1, 7, 15, 91, 95, 96–97, 129, 132, 201–202
    electric-furnace steelmaking in, 91, 96, 103, 105, 122, 123
    employment in, 7, 119, 129, 202, 224
    energy programs, and, 132
    Environmental Protection Agency, and, 102
    exchange rates, imports, steel, and, 130, 131, 204
    expansion in, 1, 16–17, 92–93, 94, 97, 105–106, 117, 221
    Experimental Negotiating Agreement in, 118
    Failing Company Doctrine, and, 113
    Fairless Works, 120
    finishing facilities in, 96, 109, 126
    Firth Sterling, 113
    Florida Steel, 122
    Ford Motor Company, steel division of, 92, 93
    formed coke in, 112
    Georgetown Steel, 111
    Granite City Steel, 91
    Great Lakes area, in, 120

greenfield-site plants, projected in, 93, 117
growth of, 92–96, 119, 194
Hamilton Works, 95
hot-strip mills of, 126
import problem of, 127–132, 197, 201–206, 233, 234
imports steel: prices of, 128–129, 130, 131; subsidized, and, 5, 97, 116–117, 128, 129, 130, 131, 133, 191, 203, 205, 211, 214
Inland Steel, 91, 93, 99, 121
integrated producers in, 91
Interlake, 92, 99
iron ore, supply of, 9, 97, 98–101
Jessop Steel, 113
Jones & Laughlin, 91, 99, 107, 113, 121
Kaiser Steel, 69, 91, 94, 107, 119, 120
Korean War, and, 92, 106
Korf mills, in, 122
labor negotiations in, 114–115, 118, 194
Laclede Steel, 122
Ling-Tempco-Vought, 113
location, geographic, of, 119–121
Lone Star Steel, 92, 113, 120
losses, financial, in, 91, 96, 97, 127, 201, 202, 203, 227
Lukens Steel, 123
Lykes Steamship Company, 113
minimills, 93, 94, 109, 119, 122–123, 221
National Steel, 91, 93, 94, 95, 99, 107, 117, 119, 121, 229
Northwest Industries, 113
Northwestern Steel and Wire, 93, 94
Nucor, 122
NVF, 113
Office of Technology Assessment, U.S., and, 110, 112
open-hearths in, 94, 105–107, 119, 121–122
operating rates in, 95, 97, 121, 127, 220
Oregon Steel Mills, 111
outlook for, 94, 112–113, 119–134, 226
ownership, private, of, 5, 113, 216
Penn-Dixie, 122
Phoenix Steel, 123
Pittsburgh area, in, 120, 121
pollution control in, 2, 97
price structure of, 5, 97, 128, 199, 201, 202, 203, 204, 205
problems, future, of, 124–133
production, steel, by, 1, 4, 17, 91, 94, 95, 106, 129, 133, 192, 194, 196, 220
Raritan River Steel, 122
raw materials for, 8, 9, 97–105
reorganization of, 91, 92, 132, 133
replacement need of, 95–96
Republic Steel, 91, 94, 99, 110, 121
Revenue Act of 1962, and, 117–118
Rouge Steel, 92

scrap for, 9, 103–105, 122, 123
Service Centers in, 124
seamless-tube mills in, 94, 123, 124
Sharon Steel, 91, 92, 113
shipments: automobile industry to, 124–125, 126; of, 97, 114, 125, 128
Sparrows Point plant, 94
steelmaking processes, future, in, 122
Steelworkers' Union, and, 118, 197
strike, 1959, in, 113, 194
support industries, imports, and, 129
technology in, 105–113
trade, international, steel, impact on, 6–7, 91, 97, 113–119, 127–132, 191, 194, 197, 201–202, 231, 232, 233, 234, 236, 237, 238
transportation, water, in, 120
trigger-price mechanism, and, 5, 7, 48, 97, 116, 129–130, 203, 204, 205, 207
Truman, President, and, 106
ultra-high power, electric furnaces, for, 112, 122
unemployment in, 127, 234
United States Steel, 85, 91, 93, 94, 95, 99, 109, 114, 117, 118, 119, 120, 121, 124, 139, 140, 143, 157, 173, 203, 207, 227, 229, 234
wage rates in, differentials, overseas, from, 131–132
water-cooled panels, electric furnace, for, 112, 122
West Coast, plants proposed for, 95
West Germany steel industry, compared to, 25
Western European steel industry, compared to, 91, 96
Wheeling-Pittsburgh, 92, 99
Wheeling Steel, 91
Wisconsin Steel, 91, 93
Youngstown Sheet and Tube, 91, 93, 113, 117, 202
United Steelworkers of America, 113, 114–115, 118, 194, 197
Usiminas-Usinas Siderurgieas de Minas Gerais SA, 161
Usinor, Union Siderurgique du Nord & de l'Est de la France, 21, 27, 28, 46, 205, 212
USSR. See Soviet Union

Venezuela:
    direct reduction, in, 11, 172
    iron ore in, 8, 37, 51, 98, 99, 100, 174, 224
    trade, international, steel, in, 173, 194, 195
    transportation, iron ore of, EEC, to, 37
Venezuela, steel industry in:
    Barquisimeto plant, 173
    capacity of, 172, 173

Caracas plant, 173
continuous casting in, 174
direct reduction in, 172
electric furnaces in, 172, 174
FIOR process in, 173
HIB process in, 173
HYL process in, 173
iron ore, supply of, 8, 172
losses, financial, in, 173
Midrex process in, 173
natural gas, use in, 173
open-hearth furnaces in, 172, 174
operating rate in, 173
ownership, public, in, 5, 172, 173
production, steel, of, 172
Sidor, 172
Sivensa, 173
trade, international, steel, in, 173
Tysland-Hole process in, 172
Zulia Steel Project, 173
Voluntary restraint agreements, U.S.,
imports, steel, 7, 51, 84, 197–198, 202,
237
Voest-Alpine, 54, 55, 149–150

Wabush Mines, 99, 140
West Germany:
coal, coking, subsidized, in, 224
coal exports of, 38, 40
coal imports of, 38,
coke and coal, steel industry for, 37–38
coking coal produced by, 37, 224
continuous casting in, 43, 44
direct-reduced iron, demand for, in, 44
exports, steel, of, 6, 193
France, iron-ore supplies from, 33, 35
government grants, steel industry, to, 229
iron ore produced in, 33
ownership, steel companies, of, in, 45–46,
49
Ruhrkohle of, 37–38
scrap exported by, 9, 40
steel industry in, 25–27, 45–46
steelmaking processes in, 42
subsidies, steel industry, of, 27, 29, 37,
50, 224
West Germany, steel industry in:
basic-oxygen steelmaking in, 42
capacity of, 25–26, 45–46, 221
continuous casting in, 43, 44
direct reduction in, 44–45
direct-reduced iron, demand for, in, 44
electric-furnace steelmaking in, 25, 41, 42
employment in, 51
Estel, 25
France, steel industry, in, compared to,
25
government, and, 29
Hoesch, 25
Japan, steel industry, in, compared to, 25

Klockner, 25
Korf organization, 25
Krupp, 25
Mannesmann, 25
mergers in, 25, 26, 27, 50–51
modernization of, 25
Neunkircher, 29
open-hearth steelmaking in, 42
ownership of, 5, 25, 45–46, 49
price increase, steel, deferral, Eurofer
II, under, 23
production, steel, in, 25, 192, 196
raw materials for, 33, 35, 37–38, 40,
44–45, 224
reorganization of, 25–27, 50
research and development in, 27
Rochling-Burbach, 25, 29
Saar, plants in, 28, 29
Salzgitter, 25, 45–46
scrap demand in, 45
subsidies, in, 27, 29, 37, 50, 224, 229
Thyssen, 25
United States, steel industry, in,
compared to, 25
USSR, steel industry, in, compared
to, 25
Western Canada Steel Ltd., 139
Western Europe:
automobile demand in, 3, 16
capital-goods demand in, 4, 17
coal imports, outlook for, 40
coking coal imported by, 37–40
continuous casting in, 43–44, 51
controversy, steel industry, of, in, 49–50
depression, steel, in, 19ff
direct reduction in, 44–45
energy crisis, impact on, 16, 50
iron-ore reserves in, 32
life-style, postwar, in, 2
non-EEC countries of, steel industries in,
53–60
outlook for, steel industry, in, 49–51
ownership, steel industry, of, in, 45–47,
49–50
price concessions, steel, in, 7, 19, 48, 49,
50, 228
profits, steel, in, 15, 18, 19
raw materials, steelmaking, prewar
supply of, 32
reorganization, steel industry, of, in,
20, 23–32, 50
scrap in, 40–41
scrap trade in, 9
steel industry, largest in, 25
steelmaking processes in, 42, 51
subsidies, steel industry, of, in, 49–51
supply-demand relationship, steel for, 28,
51
trade, international, steel, and, 47–49, 51
World War II, and, 2–3

Western European steel industry. *See*
European Economic Community
Wheeling Steel Corporation, 91
Wheeling-Pittsburgh Steel Corporation, 92,
99, 140
Wisconsin Steel Division, 91, 93
World Bank, 223–224
World War II, 2–3, 4, 6, 7–12, 31, 32–37,

41–45, 54, 97–98, 137, 139, 154
Wuhan Iron and Steel Corporation, 156

Yawata Steel, 87
Youngstown Sheet and Tube Company, 91
93, 113, 117
Yugoslavia, steel industry in, 53, 59, 60

About the Author

Rev. William T. Hogan, S.J., was graduated from Fordham College in 1939 and received the M.A. and the Ph.D. in economics from Fordham University in 1940 and 1948.

He has conducted economic studies of the steel industry and other basic, heavy industries for the past thirty years. During this time, he has authored a number of books, including *Productivity in the Blast-Furnace and Open-Hearth Segments of the Steel Industry,* the first detailed study on the subject of steel productivity; *The Development of American Heavy Industry in the Twentieth Century;* and *Depreciation Policies and Resultant Problems* (1967).

His five-volume work, *Economic History of the Iron and Steel Industry in the United States* (Lexington Books, 1971) covers industry developments from 1860 to 1971. The first of its companion studies, *The 1970s: Critical Years for Steel* was published in 1972.

In 1950, Father Hogan inaugurated Fordham University's Industrial Economics Research Institute, which has produced numerous studies dealing with economic problems of an industrial nature. He has appeared before legislative committees of both the U.S. Senate and the House of Representatives and has testified several times before the Ways and Means Committee of the House on legislation affecting depreciation charges and capital investment. He has also appeared before the Finance Committee of the Senate to testify on tax incentives for capital spending. He was a member of the Presidential Task Force on Business Taxation and a consultant to the Council of Economic Advisers to the President and the U.S. Department of Commerce.

During the past fifteen years, Father Hogan has visited most of the steel-producing facilities in the world and has delivered papers at steel conferences in France, the United Kingdom, Switzerland, Sweden, Czechoslovakia, Russia, Venezuela, Brazil, South Africa, India, the Philippines, and Japan. He is the author of numerous articles on various aspects of steel-industry economics.